Columba Sara Evelyn (Ed.)

Ferrocarril de Veracruz al Istmo

Columba Sara Evelyn (Ed.)

Ferrocarril de Veracruz al Istmo

Nationalization, Rail transport, Jesús Carranza, Veracruz

Fec Publishing

Imprint

Permission is granted to copy, distribute and/or modify this document under the terms of the GNU Free Documentation License, Version 1.2 or any later version published by the Free Software Foundation; with no Invariant Sections, with the Front-Cover Texts, and with the Back- Cover Texts. A copy of the license is included in the section entitled "GNU Free Documentation License".

All parts of this book are extracted from Wikipedia, the free encyclopedia (www.wikipedia.org).

You can get detailed informations about the authors of this collection of articles at the end of this book. The editors (Ed.) of this book are no authors. They have not modified or extended the original texts.

Pictures published in this book can be under different licences than the GNU Free Documentation License. You can get detailed informations about the authors and licences of pictures at the end of this book.

The content of this book was generated collaboratively by volunteers. Please be advised that nothing found here has necessarily been reviewed by people with the expertise required to provide you with complete, accurate or reliable information. Some information in this book maybe misleading or wrong. The Publisher does not guarantee the validity of the information found here. If you need specific advice (f.e. in fields of medical, legal, financial, or risk management questions) please contact a professional who is licensed or knowledgeable in that area.

Any brand names and product names mentioned in this book are subject to trademark, brand or patent protection and are trademarks or registered trademarks of their respective holders. The use of brand names, product names, common names, trade names, product descriptions etc. even without a particular marking in this works is in no way to be construed to mean that such names may be regarded as unrestricted in respect of trademark and brand protection legislation and could thus be used by anyone.

Cover image: www.ingimage.com
Concerning the licence of the cover image please contact ingimage.

Publisher:
Fec Publishing is a trademark of
International Book Market Service Ltd., 17 Rue Meldrum, Beau Bassin, 1713-01 Mauritius
Email: info@bookmarketservice.com
Website: www.bookmarketservice.com

Published in 2011

Printed in: U.S.A., U.K., Germany. This book was not produced in Mauritius.

ISBN: 978-613-7-70491-2

Contents

Ferrocarril de Veracruz al Istmo

The **Ferrocarril de Veracruz al Istmo** (*Vera Cruz and Isthmus Railway*) was one of the primary pre-nationalization railways of Mexico. Incorporated in West Virginia in 1898 as the **Vera Cruz and Pacific Railroad** (*Ferrocarril de Veracruz al Pacífico*), it built a line from Córdoba to Jesús Carranza on the Tehuantepec National Railway, with a branch from Veracruz to Tierra Blanca. The Mexican government gained control in May 1904 and organized the Ferrocarril de Veracruz al Istmo to operate the property under lease.[1] [2]

Control was transferred to the government-owned Ferrocarriles Nacionales de México (*National Railways of Mexico*) in 1910, and in November 1913 that company took over the property and operations.[1] [3] Following privatization in the 1990s, Ferrosur acquired the former Veracruz al Istmo.

References

[1] Fred Wilbur Powell, The Railroads of Mexico (http://books.google.com/books?id=nj8aAAAAMAAJ&pg=PA153), Stratford Company
     (Boston), 1921, pp. 153-154

[2] Manual of Statistics Company (New York), The Manual of Statistics: Stock Exchange Hand-Book (http://books.google.com/
     books?id=nzA5AAAAMAAJ&pg=RA1-PA322), 1908, p. 322

[3] Poor's Manual Company (New York), Poor's Intermediate Manual of Railroads (http://books.google.com/books?id=qqkoAAAAYAAJ&
     pg=RA1-PA928), 1917, pp. 928-946

Nationalization

Nationalisation, also spelled **nationalization**, is the process of taking an industry or assets into government ownership by a national government or state.[1] Nationalization usually refers to private assets, but may also mean assets owned by lower levels of government, such as municipalities, being transferred to the public sector to be operated by or owned by the state. The opposite of nationalization is usually privatization or de-nationalization, but may also be municipalization.

A **renationalization** occurs when state-owned assets are privatized and later nationalized again, often when a different political party or faction is in power. *A renationalization* process may also be called *reverse privatization*. Nationalization has been used to refer to either direct state-ownership and management of an enterprise or to a government acquiring a large controlling share of a nominally private, publicly listed corporation.

The motives for nationalization are political as well as economic. It is a central theme of certain fascist and/or 'state socialist' policies that the means of production, distribution and exchange, should be owned by the state on behalf of the citizenry to allow for centrally planned allocation of output, consolidation of resources,

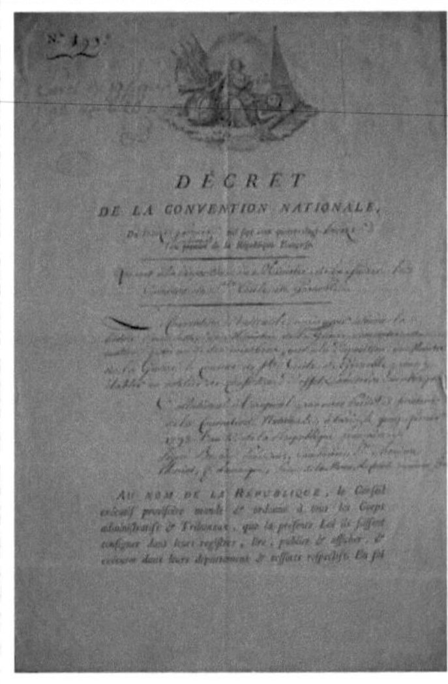

A decree of the French Revolution, assigning a convent to the army

and central planning or control of the economy. Many socialists believe that public ownership enables people to exercise full democratic control over the means whereby they earn their living and provides an effective means of distributing output to benefit the public at large, and a means for providing public finance.

Nationalized industries, charged with operating in the public interest, may be under strong political and social pressures to give much more attention to externalities. They may be obliged to operate some loss making activities where social benefits are clearly greater than social costs — for example, rural postal and transport services. As an instance, the United States Postal Service is guaranteed its nationalised status by the Constitution. The government has recognized these social obligations and, in some cases, provides subsidies for such non-commercial operations.

Since the nationalised industries are state owned, the government is responsible for meeting any debts incurred by these industries. The nationalized industries do not normally borrow from the domestic market other than for short-term borrowing. However, if they are profitable, the profit is often used as a means to finance other state services, such as social programs and government research — which can help lower the tax burden.

Nationalization may occur with or without compensation to the former owners. If it takes place without compensation it is a case of expropriation. Nationalization is distinguished from property redistribution in that the government retains control of nationalized property. Some nationalizations take place when a government seizes property acquired illegally. For example, the French government seized the car-makers Renault because its owners had collaborated with the Nazi occupiers of France.

Compensation

A key issue in nationalization is payment of compensation to the former owner. The most controversial nationalizations, known as expropriations, are those where no compensation, or an amount far below the likely market value of the nationalized assets, is paid. Many nationalizations through expropriation have come after revolutions, in particular marxist revolutions.

The traditional Western stance on compensation was expressed by United States Secretary of State Cordell Hull, during the 1938 Mexican nationalization of the petroleum industry, that compensation should be "prompt, effective and adequate." According to this view, the nationalizing state is obligated under international law to pay the deprived party the full value of the property taken. The opposing position has been taken mainly by developing countries, claiming that the question of compensation should be left entirely up to the sovereign state, in line with the Calvo Doctrine. Communist states have held that no compensation is due, based on socialist disregard for the notion of property rights.

In 1962, the United Nations General Assembly adopted Resolution 1803, "Permanent Sovereignty over National Resources", which states that in the event of nationalization, the owner "shall be paid appropriate compensation in accordance with international law." In doing so, the UN rejected both the traditional Calvo-doctrinist view and the Communist view. The term "appropriate compensation" represents a compromise between the traditional views, taking into account the need of developing countries to pursue reform even without the ability to pay full compensation, and the Western concern for protection of private property.

When nationalizing a large business, the cost of compensation is so great that many legal nationalizations have happened when firms of national importance run close to bankruptcy and can be acquired by the government for little or no money. A classic example is the UK nationalization of the British Leyland Motor Corporation. At other times, governments have considered it important to gain control of institutions of strategic economic importance, such as banks or railways, or of important industries struggling economically. The case of Rolls-Royce plc, nationalized in 1971, is an interesting blend of these two arguments. This policy was sometimes known as ensuring government control of the "commanding heights" of the economy, to enable it to manage the economy better in terms of long-term development and medium-term stability. The extent of this policy declined in the 1980s and 1990s as governments increasingly privatized industries that had been nationalized, replacing their strategic economic influence with use of the tax system and of interest rates.

Nonetheless, national and local governments have seen the advantage of keeping key strategic assets in institutions that are not strongly profit-driven and can raise funds outside the public-sector constraints, but still retain some public accountability. Examples from the last five years in the United Kingdom include the vesting of the British railway infrastructure firm Railtrack in the not-for-profit company Network Rail, and the divestment of much council housing stock to "arms-length management companies", often with mutual status.

Notable nationalizations by country

Argentina

- **1918** University Revolution and university nationalization
- **2009** Aerolíneas Argentinas renationalization

Australia

- **1948** The Australian government attempted to nationalize the banks, but the act was declared unconstitutional by the High Court of Australia in the case Bank of New South Wales v Commonwealth.[2]

Bolivia

- **2006** On May 1, 2006, newly elected Bolivian president Evo Morales announces plans to nationalize the country's natural gas industry; foreign-based companies are given six months to renegotiate their existing contracts.

Canada

- **1918** Canadian National Railways, created from several systems nationwide following their bankruptcy during and after World War I, and since privatized. (Air Canada, Canadian Broadcasting Corporation, Marine Atlantic and Via Rail were all subsidiaries of the company at one time)
- **1944** Hydro-Québec, first created through partial nationalization of electricity concerns around Montreal in Quebec by the Liberal government of Adélard Godbout. During the Quiet Revolution of the early 1960s, the remaining 11 privately owned electricity companies in Quebec were nationalized by the Liberal government of Jean Lesage.
- **1975** Potash Corporation of Saskatchewan, Province of Saskatchewan nationalized part of the potash industry. Many potash producers agreed to sell to the government instead of being nationalized.

Channel Islands

- **2003** Aurigny Air Services was bought by the States of Guernsey to keep precious routes from the island to London.

Chile

- **1972** Chilean nationalization of copper mining industry by the government of Salvador Allende.

Croatia

The HDZ government, on the break-up of Yugoslavia, nationalized private agricultural, non communist nationalized, property and rezoned it under the guise of forest statesmanship when their publicly professed agenda was to only complete the nationalization of the communists. Much of this land is in the process of being restituted and the model rethought.

Cuba

The Castro government gradually expropriated all foreign-owned private companies after the Cuban Revolution of 1959. Most of these companies were owned by U.S. corporations and individuals. Bonds at 4.5% interest over twenty years were offered to U.S. companies, but the offer was rejected by U.S. ambassador Philip Bonsal, who requested the compensation up front.[3] Only a minor amount, $1.3 million, was paid to U.S. interests before deteriorating relations ended all cooperation between the two governments.[3] The United States established a registry of claims against the Cuban government, ultimately developing files on 5,911 specific companies. The Cuban government has refused to discuss the effective and adequate compensation of U.S. claims. The United States government continues to insist on compensation for U.S. companies. In 1966-68, the Castro government nationalized all remaining privately owned business entities in Cuba, down to the level of street vendors.

Czechoslovakia

- **1945** Large manufacturing enterprises.
- **1948** All manufacturing enterprises.

Egypt

- **1956** On July 26, 1956 Egyptian President Gamal Abdel Nasser nationalized the Suez Canal Company.

France

Nationalization in France dates back to the 'regies' or state monopolies first organized under the *Ancien Régime*, for example, the monopoly on tobacco sales. Communications companies France Telecom and La Poste are relics of the state postal and telecommunications monopolies.

There was a major expansion of the nationalised sector following World War II.[4] A second wave followed in 1982.

- **1938** Societe Nationale des Chemins de Fer Francais (SNCF) (originally a 51% State holding, increased to 100% in 1982)[4]
- **1945** Several nationalizations in France, including most important banks and Renault.[4] The firm was seized for Louis Renault's alleged collaboration with Nazi Germany, although this condemnation was without judgement and after his death, making this case remarkable and rare. A later judgement (1949) admitted that Renault's plant never collaborated. Renault was successful but unprofitable whilst nationalised and remains successful today, after having been privatized in 1996.
- **1946** Charbonnages de France, Electricite de France (EdF), Gaz de France (GdF)
- **1982** A large part of the banking sector and industries of strategic importance to the state, especially in electronics and communications, were nationalized under the new president François Mitterrand and the PS-led government. Many of those companies were privatized again after 1986.

The Paris regional transport operator, Regie Autonome des Transports Parisiens (RATP), can also be counted as a nationalised industry.

Germany

The German railways were nationalised after World War I. Partial privatisation of Deutsche Bahn is currently underway, as of 2008.

Most enterprises in East Germany were nationalised following World War II. After reunification, an agency, Treuhand, was established to return them to private ownership. However, due to structural and economic problems inherent in the previous regime, many of these had to be liquidated.

- **2008** Renationalization of the "Bundesdruckerei" (Federal Print Office), which had been privatized in 2001.

Greece

- **1974** Nationalization of Olympic Airlines, main airline of Greece. The company was bought out by its founder, Aristotle Onassis.

Iceland

- **2008** Renationalization of Iceland's largest commercial banks: Kaupþing, Landsbanki, Glitnir and Icebank.
- **2009** Nationalization of Straumur Investment Bank and the savings bank SPRON.

India

The nationalised banks were credited by some, including Home minister P. Chidambaram, to have helped the Indian economy withstand the global financial crisis of 2007-2009.[5] [6]

- **1949** (1 January) Reserve Bank of India nationalised (Ref.- Reserve Bank of India chronology of events). The Reserve Bank of India was state-owned at the time of Indian independence.
- **1953** Air India under the Air Corporations Act 1953.
- **1955** Imperial Bank of India and its subsidiaries (State Bank of India and its subsidiaries)
- **1969** Nationalization of 14 Indian banks.
- **1973** Coal industry and Oil companies
- **1980** Another six banks nationalized
- 1972 Nationalisation of 106 insurance companies into four

Iran

- **1953** Iranian Prime Minister Mohammed Mossadegh nationalized the Anglo-Persian Oil Company in Iran.

Ireland

Railways in the Republic of Ireland were nationalised in the 1940s as Coras Iompair Eireann.

- **2007** On August 3, 2007, the Irish government were offered a stake in Eircom's copper network infrastructure.[7] Ireland's telephone networks were privatised in 1999.
- **2009** On January 16, 2009, the Irish Government nationalised Anglo Irish Bank to secure the bank's viability.[8]
- **2010** Irish state owned bank Anglo Irish Bank is to take majority control of one of Ireland's largest companies QUINN group bringing it under Public ownership.[9]

Israel

- **1983** Nationalization of the major Israeli banks: Bank Hapoalim, Bank Leumi, Discount Bank, Mezrachi Bank due to the Bank stock crisis that struck Israel in 1983.

Italy

- **1905** The Italian railways were nationalised as Ferrovie dello Stato.

The regime of Benito Mussolini extended nationalisation, creating the Istituto per la Ricostruzione Industriale (IRI) as a State holding company for struggling firms, including the car maker Alfa Romeo. A parallel body, Ente Nazionale Idrocarburi (Eni) was set up to manage State oil and gas interests.

Japan

- **1906** Railway Nationalization Act of Japan nationalized 17 railway companies to form the nationwide railway network that was later called Japanese National Railways.

Latvia

In **2008** Parex Bank was nationalized by Latvian governmant.

Malta

- **1974** Bank of Valletta is founded following nationalisation of the National Bank of Malta

Mexico

- **1938** The Expropriation of the Petroleum Industry of Mexico: President Lázaro Cárdenas issued a decree that the petroleum companies were in rebellion against the government of Mexico and under the powers granted him under the Expropriation Act passed by the Congress of Mexico in late 1936 expropriated them. March 19, 1938, union personnel took control of the properties.[10]
- **1982** The nationalization of the Mexican banking system made by President José López Portillo, later in the Carlos Salinas de Gortari presidency (1988–1994) a large number of banks were privatized.

The Netherlands

- **2008** The Dutch State nationalizes the Dutch activities of Belgian-Dutch banking and insurance company Fortis, which had come in solvability problems due to the international financial crisis.

New Zealand

- **2001** Central government purchased the Auckland railway network from Tranz Rail.
- **2003** The Labour Government of New Zealand took an 80% stake in near-bankrupt national air carrier Air New Zealand in exchange for a large financial infusion.
- **2004** The rest of the country's rail network is purchased from Toll New Zealand, formerly known as Tranz Rail. A new state owned enterprise, ONTRACK, was established to maintain the rail infrastructure.
- **2008** The rolling stock of Toll New Zealand was purchased by central government, bringing the rail system under total state ownership and renamed KiwiRail.

Pakistan

- **1972** On January 2, 1972, Zulfiqar Ali Bhutto, after the fall of East Pakistan, announced the nationalisation of all major industries, including iron and steel, heavy engineering, heavy electricals, petrochemicals, cement and public utilities except textiles industry and lands. [11]

Philippines

During the administration of Ferdinand Marcos, important companies such as PLDT, Philippine Airlines, Meralco and the Manila Hotel were nationalized. Other companies were sometimes absorbed into these government-owned corporations, as well as other companies, such as Napocor and the Philippine National Railways, which in their own right are monopolies (exceptions are Meralco and the Manila Hotel). Today, these companies have been reprivatized and some, such as PLDT and Philippine Airlines, have been de-monopolized. Others, like government-formed and owned Napocor, are in the process of privatization.

Poland

- **1946** Following World War II the People's Republic of Poland nationalized all enterprises with over 50 employees.

Portugal

- **1974** In the years following the Carnation Revolution, the Junta de Salvação Nacional and Provisional Governments nationalized all the banking, insurance, petrol and industrial companies. Among those companies were Companhia União Fabril (CUF), the assets of the Champalimaud family and SONAE. Along with the telecommunications companies, which were state-owned even before the Revolution, many of the nationalized companies were reprivatized in the 1980s and 1990s. In the agricultural sector, according to government estimates, about 900,000 hectares (2,200,000 acres) of agricultural land were occupied between April 1974 and December 1975 in the name of land reform; about 32% of the occupations were ruled illegal. In January 1976, the government pledged to restore the illegally occupied land to its owners, and in 1977, it promulgated the Land Reform Review Law. Restoration of illegally occupied land began in 1978.[12] [13]
- **2008**: BPN - Banco Português de Negócios bank nationalised to prevent its collapse.

Romania

- **1948** With the Decree 119 of June 11, 1948, the new Romanian communist regime nationalised all the existing private companies and their assets in Romania leading to the transformation of the Romanian economy from a market economy to a planned economy.
- **1950** With the Decree 92 of April 19, 1950, a huge number of private houses and lands are confiscated.

Russia

- **1998** The Yeltsin government began seizing Gazprom assets, claiming that the company owed back taxes. Privatization of Gazprom from the mid 1990s had been reduced to 38.37% with the intention of achieving full privatization. However, the stake of the Russian Government in Gazprom has since been increased to 50% with Vladimir Putin's plan to increase the stake to a controlling position. Gazprom is also buying up both Russian and other international utility companies.

South Korea

- **1946** USAMGIK nationalized all South Korean private railroad companies and made Department of Transportation. This now becomes Korail.

Soviet Union

- **1918** All manufacturing enterprises, many retailing enterprises, any private enterprises, the whole bank system, agrarian sector, others. Basically everything was *nationalized* in the name of the Revolution (a justification phrased used upon the nationalization process), no private ownership was allowed. Later the government of Lenin introduced the New Economic Policy that slightly reverted process, however upon the death of the Soviet *vozhd* (Lenin) Stalin renewed the process again.

Spain

- **1941** Spain's railways were nationalised, as RENFE, in the aftermath of the Spanish Civil War.
- **1983** Nationalization without compensation of the Spanish Rumasa. Separate business were later privatized.

Sri Lanka

- **1958** The Government nationalised Bus transport (creating the Ceylon Transport Board). The Colombo Port was also nationalised the same year.
- **1961** The local subsidiaries of the foreign owned petroleum companies, Caltex, Esso and Shell had formed a cartel, to break which they were nationalised. The Insurance companies and the Bank of Ceylon were also nationalised in the same year.
- **1971** Graphite mines nationalised.
- **1972** Locally owned Tea and Rubber plantations were nationalised under the Land Reform law.
- **1975** *Sterling* plantation companies (owned by British plantation companies) were nationalised.
- **2009** Seylan Bank nationalised to prevent its collapse.

Sweden

- **1939-1948** Nationalisation of most of the private Railway companies.
- **1957** The mining company LKAB is nationalized. The state had owned 50% of the corporation's shares, with options to buy the remainder, since 1907.[14]
- **1992** A large part of Sweden's banking sector is nationalized.[15]

United Kingdom

The following companies/industries were the subject of nationalisation in the given year:

- **1868** Nationalisation of inland telegraphs under the GPO[16]
- **1875** Suez Canal Company - The Egyptian share in the company was bought out by the British Government.
- **1912** Nationalisation of inland telephone services under the GPO, apart from Portsmouth, Hull, Guernsey, and Jersey. The Portsmouth telephone service was nationalised the following year.
- **1916** Liquor Trade - The nationalisation of pubs and breweries in Carlisle, Gretna, Cromarty and Enfield under the State Management Scheme; mainly an attempt to restricting alcohol consumption by armaments factory workers. The scheme was privatised by asset transfer in 1973.[17]
- **1926** Central Electricity Board introduced under **The Electricity (Supply) Act 1926** established the National Grid and set up a national standard for electricity supply in the UK.
- **1927** British Broadcasting Company (a privately owned company) became British Broadcasting Corporation (BBC), a public corporation operating under a Royal Charter.
- **1933** London Transport
- **1938** Nationalisation of UK Coal Royalties under the Coal Commission[18]
- **1939** British Overseas Airways Corporation (BOAC) later to become British Airways (BA) - combining the private British Airways Ltd. and the state owned Imperial Airways
- **1939** At the outset of World War II, much of British industry was subjected to State regulation or control, although not nationalised as such.
- **1943** North of Scotland Hydro-Electricity Board
- **1946** Coal industry under the National Coal Board (later British Coal); Bank of England - the latter had had private shareholders who were bought out by the state.
- **1947** Central Electricity Generating Board and area electricity boards, Cable & Wireless Ltd - the latter had had private shareholders who were bought out by the state.

- **1948** National rail, inland (not marine) water transport, some road haulage, some road passenger transport and Thomas Cook & Son under the British Transport Commission. Separate elements operated as British Railways, British Road Services, and British Waterways, also national health services created (as England and Wales, for Scotland and for Northern Ireland) taking over a mixture of previously local authority, private commercial and charitable organisations.
- **1949** Local authority gas supply undertakings in England, Scotland and Wales
- **1951** Iron and Steel Industry (denationalised by the following Conservative Government)[19]
- **1967** British Steel
- **1969** National Bus Company, combining former interests of the British Transport Commission with others acquired from the British Electric Traction group.
- **1969** Post Office Corporation created
- **1971** Rolls-Royce (1971) Ltd - The strategically important aero-engine part of the recently bankrupt Rolls Royce Limited.
- **1973** Local authority water supply undertakings in England and Wales
- **1973** British Gas plc Corporation created, replacing regional gas boards.
- **1974** British Petroleum - the combination of a 50% stake bought by Winston Churchill as First Lord of the Admiralty after World War I with around a 25% stake acquired by the Bank of England from Burmah Oil made the UK Government directly or indirectly BP's majority shareholder, though commercial independence was maintained. The shares were all sold during the 1980s.
- **1975** National Enterprise Board - a State holding company for full or partial ownership of industrial undertakings
- **1976** British Leyland Motor Corporation - became *British Leyland* upon nationalization. Privatized in 1986 to British Aerospace.
- **1977** British Aerospace - combining the major aircraft companies British Aircraft Corporation, Hawker Siddeley and others. British Shipbuilders - combining the major shipbuilding companies including Cammell Laird, Govan Shipbuilders, Swan Hunter, Yarrow Shipbuilders
- **1981** British Telecom created, taking control of telecommunications services from the Post Office
- **1984** Johnson Matthey - purchased for a nominal sum of £1 by the Thatcher government [20]
- **1997** Docklands Light Railway - John Prescott announced to the 1997 Labour Party Conference that he had nationalised this, although it was already in public hands anyway.[20]
- **2001** Railtrack - The owner and operator of the railway infrastructure, Railtrack, was not nationalised as such. However, its replacement Network Rail, whilst not a state-owned company, has no shareholders *(company limited by guarantee)* and is underwritten by the state. In addition prior to this the government began to make use of a residual shareholding of 0.2% (including voting rights) in Railtrack Group Plc leftover from the original sale.[21]
- **2008** Northern Rock - announced by Alistair Darling, Chancellor of the Exchequer on 17 February 2008 as 'a temporary measure'. The bank will be run at 'arms length' as a commercial business and sold to a private buyer in the future.[22]
- **2008** Bradford & Bingley (mortgage book only) - announced by Alistair Darling, Chancellor of the Exchequer on 29 September 2008. The loans part of the company was nationalised, while the commercial bank was sold off.[23]
- **2008** In October, the Royal Bank of Scotland, and the newly merged HBOS-Lloyds TSB was partly nationalised. The Government took over approximately 60% of RBS (later increased to 70%, then 80%) and 40% of HBOS-Lloyds TSB. This is part of the £500bn bank rescue package.
- **2009** On 13 November, Directly Operated Railways, a government company, took over the East Coast Main Line railway franchise that National Express had bought in 2007 for £1.4 billion, a sum originally to be paid over 7 years. The nationalised service operates as East Coast and includes services from London to York and Edinburgh. It has been stated by the government that their control is a temporary measure, initially to last 2 years.

British assets nationalised by other countries

- **1940s** Argentine railways
- **1953** British Petroleum's Iranian assets by their government (actually a nationalisation of part of a part-nationalised company)
- **1956** The Egyptian Government nationalised the Suez Canal, owned by the Suez Canal Company which was part owned by the British State.
- **1962** The Sri Lanka Government nationalised the assets in the country of the partly British-owned Royal Dutch Shell company.
- **1975** The Sri Lanka Government nationalised the assets in the country of the British-owned plantation companies.

United States

- **1862**: The Legal Tender Act nationalized the monetary system under fiat currency.
- **1863**: The National Bank Act nationalized the banking system and further monopolized the money supply.
- **1917**: All U.S. railroads were nationalized as the Railroad Administration during World War I as a wartime measure. The United States Railroad Administration was returned to private ownership in 1920.
- **1939**: Organization of the Tennessee Valley Authority entailed the nationalization of the facilities of the former Tennessee Electric Power Company.
- **1971**: The National Railroad Passenger Corporation (Amtrak) is a government-owned corporation created in 1971 for the express purpose of relieving American railroads of their legal obligation to provide inter-city passenger rail service. The (primarily) freight railroads had petitioned to abandon passenger service repeatedly in the decades leading up to Amtrak's formation.
- **1976**: The Consolidated Rail Corporation (Conrail), another government corporation, was created to take over the operations of six bankrupt rail lines operating primarily in the Northeast; Conrail was privatized in 1987. Initial plans for Conrail would have made it a truly nationalized system like that during World War I, but an alternate proposal by the Association of American Railroads won out.
- **1980s**: Resolution Trust Corporation seized control of hundreds of failed Savings & Loans.
- **2001**: In response to the September 11 attacks, the then-private airport security industry was nationalized and put under the authority of the Transportation Security Administration.
- **2008**: Some economists consider the U.S. government's takeover of the Federal Home Loan Mortgage Corporation and Federal National Mortgage Association to have been nationalization.[24] [25] The conservatorship model used with Fannie Mae and Freddie Mac is looser and more temporary than nationalization.[26]
- **2009**: Some economists consider the U.S. government's actions with regards to Citigroup to have been a partial nationalization.[27] Proposal was made that banks like Citigroup be brought under a conservatorship model similar to Fannie Mae and Freddie Mac, that some of their "good assets" be dropped into newly created "good bank" subsidiaries (presumably under new management), and the remaining "bad assets" be left to be managed under the supervision of a conservatorship structure.[26] The U.S. government's actions with regard to General Motors in replacing the CEO with a government approved CEO is likewise being considered as nationalization.[28] [29] On June 1, 2009, General Motors filed for bankruptcy, with the United States investing up to $50 billion and taking 60% ownership in the company. President Obama stated that the nationalization was temporary, saying, "We are acting as reluctant shareholders because that is the only way to help GM succeed"[30]

Venezuela

- **2007** On May 1, 2007, Venezuela stripped the world's biggest oil companies of operational control over massive Orinoco Belt crude projects, a controversial component in President Hugo Chavez's nationalization drive.
- **2008** On April 3, 2008, President Hugo Chavez ordered the nationalization of the cement industry.[31]
- **2008** On April 9, 2008, Hugo Chavez ordered the nationalization of Venezuelan steel mill Sidor, in which Luxembourg-based Ternium currently holds a 60% stake. Sidor employees and the Government hold a 20% stake respectively.[32]
- **2008** On August 19, 2008, Hugo Chavez ordered the take-over of a cement plant owned and operated by Cemex, an international cement producer. While shares of Cemex fell on the New York Stock Exchange, the cement plant comprises only about 5% of the company's business, and is not expected to adversely affect the company's ability to produce in other markets. Chavez has been looking to nationalize the concrete and steel industries of his country to meet home building and infrastructure goals.[33]
- **2009** On February 28, 2009, Hugo Chavez ordered the army to take over all rice processing and packaging plants.[34]
- **2010** On January 20, 2010, Hugo Chavez signed an ordinance to nationalize six supermarkets in Venezuela under the system of retail stores of a French company because of increasing price and speculation hoarding illicit.[35]
- **2010** On June 24, 2010, Venezuela announced the intention to nationalize oil drilling rigs belonging to the U.S. company Helmerich & Payne.[36]
- **2010** On October 25, 2010, Chavez announced that the government was nationalizing two U.S.-owned Owens-Illinois glass-manufacturing plants.[37]
- **2010** On October 31, 2010, Venezuelan President Hugo Chavez said his government will take over the Sidetur steel manufacturing plant. Sidetur is owned by Vivencia, which had two mineral plants appropriated by the government in 2008.[37]

Zimbabwe

- Zimbabwe has nationalized its food distribution infrastructure.
- Zimbabwe Cricket formerly the Zimbabwe Cricket Union was nationalised in 2004

Other countries

- Nationalization of the oil industry in numerous countries, including Libya, Kuwait, Mexico, Nigeria, Saudi Arabia, and Venezuela.

See also

- Compulsory purchase
- Constitutional economics
- Eminent domain
- Confiscation
- Government agency
- Planned economy
- Privatization - the reverse process
- Public ownership
- Railway nationalization
- Reprivatization
- Sequestration
- State capitalism
- State-owned enterprise

- State sector
- Socialization (economics) - the process of making cooperative in management or ownership

References

[1] http://www.merriam-webster.com/dictionary/nationalization

[2] The Constitutional Centre of Western Australia | The Role of The High Court (http://www.ccentre.wa.gov.au/ForSchools/Backgroundinformation/TheThreeArmsofGovernment/Pages/TheRoleoftheHighCourt.aspx)

[3] Thomas, Hugh (March 1971). *Cuba; the Pursuit of Freedom*. New York: Harper & Row. pp. 224, p252. ISBN 0060142596.

[4] Myers (1949)

[5] PSU banks' policies saved India from financial blushes: Chidambaram (http://economictimes.indiatimes.com/News/Economy/Policy/PSU_banks_policies_saved_India_from_financial_blushes_Chidambaram/rssarticleshow/3901203.cms)

[6] The importance of public banking (http://www.hinduonnet.com/fline/fl2525/stories/20081219252504800.htm)

[7] Eircom and State in broadband swap? (http://www.rte.ie/news/2007/0803/eircom-business.html)

[8] Government nationalises 'fragile' Anglo Irish Bank (http://www.irishtimes.com/newspaper/frontpage/2009/0116/1232059654021.html)

[9] Anglo Irish Bank's €700m Quinn plan (http://www.rte.ie/news/2010/0408/quinn-business.html)

[10] The Expropriation of the Petroleum Industry of Mexico in 1938 (http://www.sjsu.edu/faculty/watkins/pemex3.htm)

[11] US Country Studies. "Zulfikar Ali Bhutto" (http://countrystudies.us/pakistan/20.htm) (PHP). . Retrieved 2006-11-07.

[12] "Portugal" (http://countrystudies.us/portugal/63.htm), *Country Studies* (U.S. Library of Congress), , "In the mid-1980s, agricultural productivity was half that of the levels in Greece and Spain and a quarter of the EC average. The land tenure system was polarized between two extremes: small and fragmented family farms in the north and large collective farms in the south that proved incapable of modernizing. The decollectivization of agriculture, which began in modest form in the late 1970s and accelerated in the late 1980s, promised to increase the efficiency of human and land resources in the south during the 1990s."

[13] "Portugal Agriculture" (http://www.nationsencyclopedia.com/Europe/Portugal-AGRICULTURE.html), *The Encyclopedia of the Nations*,

[14] A Historic Journey (http://www.lkab.com/__C12570A1002EAAAE.nsf/($all)/986496EB999E5F43C1257165002ECDCF/$file/A historic journey.pdf) Luossavaara-Kiirunavaara Aktiebolag, April 2006

[15] Stopping a Financial Crisis, the Swedish Way (http://www.nytimes.com/2008/09/23/business/worldbusiness/23krona.html)

[16] Schifferes, Steve (February 18, 2008). "The lessons of nationalisation" (http://news.bbc.co.uk/1/hi/business/7250252.stm). *BBC News*. . Retrieved May 20, 2010.

[17] http://www.historytoday.com/dt_main_allatonce.asp?gid=9859&g9859=x&g9857=x&g30026=x&g20991=x&g21010=x&g19965=x&g19963=x&amid=9859

[18] SN 1825 -Nationalisation of the UK Coal Royalties, 1938 : Compensation Payments (http://www.data-archive.ac.uk/findingData/snDescription.asp?sn=1825)

[19] http://www.uksteel.org.uk/history.htm

[20] "What was the last nationalisation?", *BBC News*, 18 February 2008 (http://news.bbc.co.uk/1/hi/magazine/7250668.stm)

[21] House of Commons Hansard Written Answers for 12 Feb 2002 (pt 16) (http://www.publications.parliament.uk/pa/cm200102/cmhansrd/vo020212/text/20212w16.htm)

[22] "Northern Rock to be nationalised" (http://news.bbc.co.uk/1/hi/business/7249575.stm). *BBC News*. February 17, 2008. . Retrieved May 20, 2010.

[23] "HIGHLIGHTS-Britain nationalises Bradford & Bingley" (http://www.reuters.com/article/economicNews/idUSBINGLEY20080929). *Reuters*. September 29, 2008. .

[24] US rescue of Fannie, Freddie poses taxpayer risks (http://www.usatoday.com/news/washington/2008-09-08-127573146_x.htm)

[25] Diamond and Kashyap on the Recent Financial Upheavals (http://freakonomics.blogs.nytimes.com/2008/09/18/diamond-and-kashyap-on-the-recent-financial-upheavals/)

[26] Baxter, Lawrence; Brown, Bill; Cox, Jim (February 27, 2009). "Finally, A Bridge to Somewhere" (http://www.huffingtonpost.com/lawrence-baxter-bill-brown-and-james-cox/finally-a-bridge-to-somew_b_170688.html). *Huffington Post*. .

[27] Nature of Citi stake debatable (http://www.ft.com/cms/s/0/df51e47e-04fa-11de-8166-000077b07658.html)

[28] Am I the Last Capitalist? Obama Falters on Rick Wagoner, GM, and the Auto Industry - Mary Kate Cary (usnews.com) (http://www.usnews.com/blogs/mary-kate-cary/2009/03/30/am-i-the-last-capitalist-obama-falters-on-rick-wagoner-gm-and-the-auto-industry-.html)

[29] "If, in fact, Wagoner resigned because somebody in government said, 'You have to resign,' then I think we have nationalized the auto industry, at least GM, and I think that's bad to have the government have a socialized car industry," -Sen. Chuck Grassley (R-Iowa) (http://www.cbsnews.com/stories/2009/03/30/politics/politico/main4905219.shtml/)

[30] *The Washington Post*. http://www.washingtonpost.com/wp-dyn/content/article/2009/06/01/AR2009060101480.html.

[31] Al Jazeera English - Americas - Chavez nationalises cement industry (http://english.aljazeera.net/NR/exeres/78BD5E2C-6A4B-4787-BAF7-7F764A8BF7A0.htm)

[32] "Venezuela to nationalize steelmaker Sidor: union" (http://www.reuters.com/article/worldNews/idUSN0942912020080409?feedType=RSS&feedName=worldNews). *Reuters*. April 9, 2008. .

[33] "Venezuela Seizes Cemex - Forbes.com" (http://www.forbes.com/markets/2008/08/19/
      cemex-venezuela-chavez-markets-equity-cx_ra_0819markets41.html). .
[34] "Chavez sends army to rice plants" (http://news.bbc.co.uk/2/hi/americas/7917176.stm). *BBC News*. March 1, 2009. . Retrieved May
      20, 2010.
[35] Venezuela quốc hữu hóa 6 siêu thị ngoại quốc (http://www.nguoi-viet.com/absolutenm/anmviewer.asp?a=107211&z=5)
      (Vietnamese)
[36] Frank Jack Daniel (June 24, 2010). "Venezuela to nationalize U.S. firm's oil rigs" (http://www.reuters.com/article/
      idUSTRE65N0UM20100624). *Reuters*. .
[37] the CNN Wire Staff (November 2, 2010). "Venezuela nationalizes private steel plant" (http://www.cnn.com/2010/WORLD/americas/
      11/01/venezuela.nationalization/index.html). *CNN.com*. .

Bibliography

On banks nationalization

- Dougherty, Carter, *Stopping a Financial Crisis, the Swedish Way* (http://www.nytimes.com/2008/09/23/
  business/worldbusiness/23krona.html), The New York Times," September 23, 2008.

- Hilferding, Rudolf (1981) *Finance Capital: A Study of the Latest Phase of Capitalist Development* (http://books.
  google.com/books?id=Cq6ZAAAAIAAJ) (London: Routledge & Kegan Paul), p. 234. ISBN 0710006187,
  9780710006189

- La Porta, Rafael, Florencio Lopez-de-Silanes, Andrei Shleifer, *Government Ownership of Banks* (http://ideas.
  repec.org/a/bla/jfinan/v57y2002i1p265-301.html), The Journal of Finance, vol. 57, No. 1 (Feb. 2002),
  265-301.

- La Botz, Dan (2008) *The Financial Crisis: Will the U.S. Nationalize the Banks?* (http://www.monthlyreview.
  org/mrzine/labotz280908.html) Monthly Review 28 September 2008

- Lohr, Steve, *From Japan's Slump in 1990s, Lessons for U.S.* (http://www.nytimes.com/2008/02/09/business/
  worldbusiness/09japan.html), The New York Times, February 9, 2008.

- Maxfield, Sylvia, *The International Political Economy of Bank Nationalization: Mexico in Comparative
  Perspective* (http://www.jstor.org/pss/2503718?cookieSet=1), *Latin American Research Review*, Vol. 27, No.
  1 (1992), pp. 75–103.

- Myers, Margaret G., *The Nationalization of Banks in France* (http://www.jstor.org/pss/2144223), *Political
  Science Quarterly*, Vol. 64, No. 2 (June

External links

- The importance of public banking (http://www.hinduonnet.com/fline/fl2525/stories/20081219252504800.
  htm) - article on Indian public sector banks
- Time for Permanent Nationalization (http://www.dollarsandsense.org/archives/2009/0309moseley.html) by
  economist Fred Moseley in Dollars & Sense magazine, January/February 2009
- The Corporate Governance of Banks - a concise discussion of concepts and evidence (http://ideas.repec.org/p/
  wbk/wbrwps/3404.html)

Rail transport

<table>
<tr><td align="center">Public Infrastructure</td></tr>
<tr><td align="center"></td></tr>
<tr><td align="center">Assets and Facilities</td></tr>
<tr><td align="center">Airport • Bridge • Broadband • Canal
Critical • Dams • Electricity
Energy • Freight • Hazardous waste
Hospitals • Levees • Lighthouses • Parks
Port • Mass transit • Public housing
Public schools • Public Space • Rail • Road
Sewage • Shipment • Solid Waste
Telecommunications
Utilities • Water locks
Water system • Wastewater</td></tr>
<tr><td align="center">Concepts</td></tr>
<tr><td align="center">Asset Management • Appropriations
Bank • Benefit tax • Build-Operate-Transfer
Design-Build • Earmark • Fixed cost
Engineering Contracts • Externality
Government debt • Life cycle assessment
Logistics • Maintenance • Monopoly
Project Management • Property Tax
Public-private partnerships • Public capital
Public finance • Public good • Public sector
Renovation • Replacement
Spillover effect • Supply chain • Taxation</td></tr>
<tr><td align="center">Ideas and Issues</td></tr>
<tr><td align="center">Air traffic control • Brownfield • Bus rapid
Carbon footprint • Containerization
Congestion pricing • Ethanol fuel
Gas Tax • Groundwater • High speed rail
Highway Trust Fund • Hybrid vehicles
Just-in-time (business) • Land use planning
Mobile data terminal • Pork barrel
Program thermostat • Recycle • Renewable
Reverse osmosis • Smart Grid
Smart Growth • Stormwater • Urban sprawl
Traffic congestion • TOD
Vehicle efficiency • Waste-to-energy
Weatherization • Wireless technology</td></tr>
<tr><td align="center">Fields of Study</td></tr>
</table>

Architecture • Civil, Electrical,
Mechanical engineering • Public economics
Public policy • Urban planning

Infrastructure Portal

Rail transport is a means of conveyance of passengers and goods by way of wheeled vehicles running on rail tracks. In contrast to road transport, where vehicles merely run on a prepared surface, rail vehicles are also directionally guided by the tracks they run on. Track usually consists of steel rails installed on sleepers/ties and ballast, on which the rolling stock, usually fitted with metal wheels, moves. However, other variations are also possible, such as slab track where the rails are fastened to a concrete foundation resting on a prepared subsurface.

Rolling stock in railway transport systems generally has lower frictional resistance when compared with highway vehicles, and the passenger and freight cars (carriages and wagons) can be coupled into longer trains. The operation is carried out by a railway company, providing transport between train stations or freight customer facilities. Power is provided by locomotives which either draw electrical power from a railway electrification system or produce their own power, usually by diesel engines. Most tracks are accompanied by a signalling system. Railways are a safe land transport system when compared to other forms of transport.[1] Railway transport is capable of high levels of passenger and cargo utilization and energy efficiency, but is often less flexible and more capital-intensive than highway transport is, when lower traffic levels are considered.

The oldest, man-hauled railways date to the 6th century B.C, with Periander, one of the Seven Sages of Greece, credited with its invention. With the British development of the steam engine, it was possible to construct mainline railways, which were a key component of the industrial revolution. Also, railways reduced the costs of shipping, and allowed for fewer lost goods. The change from canals to railways allowed for "national markets" in which prices varied very little from city to city. Studies have shown that the invention and development of the railway in Europe was one of the most important technological inventions of the late 19th century for the United States, without which, GDP would have been lower by 7.0% in 1890. In the 1880s, electrified trains were introduced, and also the first tramways and rapid transit systems came into being. Starting during the 1940s, the non-electrified railways in most countries had their steam locomotives replaced by diesel-electric locomotives, with the process being almost complete by 2000. During the 1960s, electrified high-speed railway systems were introduced in Japan and a few other countries. Other forms of guided ground transport outside the traditional railway definitions, such as monorail or maglev, have been tried but have seen limited use.

History

The history of the growth, decline and resurgence of rail transport can be divided up into several discrete periods defined by the principal means of motive power used.

Pre-steam

The earliest evidence of a railway was a 6-kilometre (3.7 mi) Diolkos wagonway, which transported boats across the Corinth isthmus in Greece during the 6th century BC. Trucks pushed by slaves ran in grooves in limestone, which provided the track element. The Diolkos ran for over 600 years.[2]

Railways began reappearing in Europe after the Dark Ages. The earliest known record of a railway in Europe from this period is a stained-glass window in the Minster of Freiburg im Breisgau in Germany, dating from around 1350.[3] In 1515, Cardinal Matthäus Lang wrote a description of the Reisszug, a funicular railway at the Hohensalzburg Castle in Austria. The line originally used

Horsecar in Brno, Czech Republic

wooden rails and a hemp haulage rope, and was operated by human or animal power. The line still exists, albeit in updated form, and is one of the oldest railways still to operate.[4] [5]

By 1550, narrow gauge railways with wooden rails were common in mines in Europe.[6] By the 17th century, wooden wagonways were common in the United Kingdom for transporting coal from mines to canal wharfs for transshipment to boats. The world's oldest working railway, built in 1758, is the Middleton Railway in Leeds. In 1764, the first gravity railroad in the United States was built in Lewiston, New York.[7] The first permanent tramway was the Leiper Railroad in 1810.[8]

The first iron plate rail way made with cast iron plates on top of wooden rails, was taken into use in 1768.[9] This allowed a variation of gauge to be used. At first only balloon loops could be used for turning, but later, movable points were taken into use that allowed for switching.[10] From the 1790s, iron edge rails began to appear in the United Kingdom.[11] In 1803, William Jessop opened the Surrey Iron Railway in south London, arguably the world's first horse-drawn public railway.[12] The invention of the wrought iron rail by John Birkinshaw in 1820 allowed the short, brittle, and often uneven, cast iron rails to be extended to 15 feet (4.6 m) lengths.[13] These were succeeded by steel in 1857.[11]

The railroad era in the United States began in 1830 when Peter Cooper's locomotive, *Tom Thumb*, first steamed along 13 miles (21 km) of Baltimore and Ohio railroad track.[14] In 1833 the nation's second railroad ran 136 miles (219 km) from Charleston to Hamburg in South Carolina.[15] Not until the 1850s, though, did railroads offer long distance service at reasonable rates. A journey from Philadelphia to Charleston involved eight different gauges, which meant that passengers and freight had to change trains seven times. Only at places like Bowling Green, Kentucky, the railroads were connected to one another.

Age of steam

The development of the steam engine during the Industrial revolution in the United Kingdom spurred ideas for mobile steam locomotives that could haul trains on tracks. James Watt's patented steam engines of 1769 (revised in 1782) were heavy low-pressure engines which were not suitable for use in locomotives. However, in 1804, using high-pressure steam, Richard Trevithick demonstrated the first locomotive-hauled train in Merthyr Tydfil, United Kingdom.[16] [17]

A British steam locomotive-hauled train

Accompanied with Andrew Vivian, it ran with mixed success,[18] breaking some of the brittle cast-iron plates.[19] Two years later, the first passenger horse-drawn railway was opened nearby between Swansea and Mumbles.[20]

The earliest British steam railways

In 1811, John Blenkinsop designed the first successful and practical railway locomotive[21] —a rack railway worked by a steam locomotive between Middleton Colliery and Leeds on the Middleton Railway. The locomotive, *Salamanca*, was built the following year.[22] :20 In 1825, George Stephenson built the *Locomotion* for the Stockton and Darlington Railway, north east England, which was the first public steam railway in the world. In 1829, he built *The Rocket* which was entered in and won the Rainhill Trials. This success led to Stephenson establishing his company as the pre-eminent builder of steam locomotives used on railways in the United Kingdom, the United States and much of Europe.[22] :24–30

In 1830, the first intercity railway, the Liverpool and Manchester Railway, opened. The gauge was that used for the early wagonways and had been adopted for the Stockton and Darlington Railway.[23] The 1435 mm (4 ft 8 $\frac{1}{2}$ in) width became known as the international standard gauge, used by about 60% of the world's railways. This spurred the spread of rail transport outside the UK.

By the early 1850s Britain had over 7,000 miles of railway, 'a stunning achievment given that only twenty years had elapsed since the opening of the Liverpool and Manchester Railway.[19]

Early railroads in the USA

Railroads (as they were known in the USA) were built on a far larger scale than those in Continental Europe, both in terms of the distances covered and also in the loading gauge adopted which allowed for heavier locomotives and double-decker trains.

The Baltimore and Ohio that opened in 1830 was the first to evolve from a single line to a network in the United States.[24] By 1831, a steam railway connected Albany and Schenectady, New York, a distance of 16 miles, which was covered in 40 minutes.[25]

The years between 1850 and 1890 saw phenomenal growth in the US railroad system, which at its peak constituted one third of the world's total mileage.[26] Although the American Civil War placed a temporary halt to major new developments the conflict did demonstrate the enormous strategic importance of railways at times of war. After the war major developments include the first elevated railway built in New York in 1867 and the symbolically important first transcontinental railway was completed in 1869.[27]

Electrification and dieselisation

Experiments with electrical railways were started by Robert Davidson in 1838. He completed a battery-powered carriage capable of 6.4 km/h (4 mph). The Giant's Causeway Tramway was the first to use electricity fed to the trains en-route, using a third rail, when it opened in 1883. Overhead wires were taken into use in 1888. At first, this was taken into use on tramways that, until then, had been horse-drawn tramcars. The first conventional electrified railway was the Roslag Line in Sweden. During the 1890s, many large cities, such as London, Paris and New York used the new technology to build rapid transit for urban commuting. In smaller cities, tramways became common and were often the only mode of public transport until the introduction of buses in the 1920s. In North America, interurbans became a common mode to reach suburban areas. At first, all electric railways used direct current but, in 1904, the Stubaital Line in Austria opened with alternating current.[28]

0-Series Shinkansen, introduced in 1964, triggered the intercity train travel boom.

Steam locomotives require large pools of labour to clean, load, maintain and run. After World War II, dramatically increased labour costs in developed countries made steam an increasingly costly form of motive power. At the same time, the war had forced improvements in internal combustion engine technology that made diesel locomotives cheaper and more powerful. This caused many railway companies to initiate programmes to convert all unelectrified sections from steam to diesel locomotion.

Elevated section of the Chicago 'L'

Following the large-scale construction of motorways after the war, rail transport became less popular for commuting and air transport started taking large market shares from long-haul passenger trains. Most tramways were either replaced by rapid transits or buses, while high transshipment costs caused short-haul freight trains to become uncompetitive. The 1973 oil crisis led to a change of mind set and most tram systems that had survived into the 1970s remain today. At the same time, containerization allowed freight trains to become more competitive and participate in intermodal freight transport. With the 1964 introduction of the Shinkansen high-speed rail in Japan, trains

Luas in Dublin, Ireland

could again have a dominant position on intercity travel. During the 1970s, the introduction of automated rapid transit systems allowed cheaper operation. The 1990s saw an increased focus on accessibility and low-floor trains. Many tramways have been upgraded to light rail and many cities that closed their old tramways have reopened new light railway systems.

Innovations

Many benchmarks in equipment and infrastructure led to the growing use of railways. Some innovative features taking place in the 19th and 20th centuries included wood cars replaced with all-steel cars, which provided better safety and maintenance; iron rails replaced with steel rails, which provided higher speed and capacity with lower weight and cost; stove-heated cars to steam-heating cars, piped from locomotive; gas lighting to electric lighting, with use of battery/alternator unit beneath the car; development of air-conditioning with additional underbody equipment and ice compartment. Some innovative rolling stock included the lightweight, diesel-powered streamliner, which was an modernistic, aerodynamically-styled train with flowing contours; then came the ultra-lightweight car with internal combustion engine in each train's power car; others included the dome car, turbined-powered trains, bilevel rolling stock, and the high-tech/high-speed electric trains.[29]

Even more, in the first half of the 20th century, infrastructure elements adopted technological changes including the continuously welded rail that was 1/4 miles long; concrete tie usage; double tracking major lines; intermodal terminal and handling technology; advances in diesel-electric propulsion to include AC traction systems and propulsion braking systems; and just-in-time inventory control. Beyond technology, even management of systems seen improvements with the adoption of environmental impact concerns; heightened concern of employee and public safety; introduction of urban area rail networks and public agencies to manage them; and downsizing of the industry employment with greater use of contractors and consultants.[30]

Trains

A train is a connected series of rail vehicles that move along the track. Propulsion for the train is provided by a separate locomotive or from individual motors in self-propelled multiple units. Most trains carry a revenue load, although non-revenue cars exist for the railway's own use, such as for maintenance-of-way purposes. The engine driver controls the locomotive or other power cars, although people movers and some rapid transits are driverless.

Haulage

Traditionally, trains are pulled using a locomotive. This involved a single or multiple powered vehicles being located at the front of the train and providing sufficient adhesion to haul the weight of the full train. This remains dominant for freight trains and is often used for passenger trains. A push-pull train has the end passenger car equipped with a driver's cab so the engine driver can remotely control the locomotive. This allows one of the locomotive-hauled trains drawbacks to be removed, since the locomotive need not be moved to the end of the train each time the train changes direction. A railroad car is a vehicle used for the haulage of either passengers or freight.

Russian 2TE10U diesel locomotive

A multiple unit has powered wheels throughout the whole train. These are used for rapid transit and tram systems, as well as many both short- and long-haul passenger trains. A railcar is a single, self-powered car. Multiple units have a driver's cab at each end of the unit and were developed following the ability to build electric motors and engines small enough to build under the coach. There are only a few freight multiple units, most of which are high-speed post trains.

Motive power

Steam locomotives are locomotives with a steam engine that provides adhesion. Coal, petroleum, or wood is burned in a firebox. The heat boils water in the fire-tube boiler to create pressurized steam. The steam travels through the smokebox before leaving via the chimney. In the process, it powers a piston that transmits power directly through a connecting rod (US: main rod) and a crankpin (US: wristpin) on the driving wheel (US main driver) or to a crank on a driving axle. Steam locomotives have been phased out in most parts of the world for economical and safety reasons although many are preserved in working order by heritage railways.

A RegioSwinger multiple unit of the Croatian Railways

Electric locomotives draw power from a stationary source via an overhead wire or third rail. Some also or instead use a battery. A transformer in the locomotive converts the high voltage, low current power to low voltage, high current used in the electric motors that power the wheels. Modern locomotives use three-phase AC induction motors. Electric locomotives are the most powerful traction. They are also the cheapest to run and provide less noise and no local air pollution. However, they require high capital investments both for the overhead lines and the supporting infrastructure. Accordingly, electric traction is used on urban systems, lines with high traffic and for high-speed rail.

Diesel locomotives use a diesel engine as the prime mover. The energy transmission may be either diesel-electric, diesel-mechanical or diesel-hydraulic but diesel-electric is dominant. Electro-diesel locomotives are built to run as diesel-electric on unelectrified sections and as electric locomotives on electrified sections.

Alternative methods of motive power include magnetic levitation, horse-drawn, cable, gravity, pneumatics and gas turbine.

Passenger trains

A passenger train travels between stations where passengers may embark and disembark. The oversight of the train is the duty of a guard/train manager. Passenger trains are part of public transport and often make up the stem of the service, with buses feeding to stations. Passenger trains can involve a variety of functions including long distance intercity travel, daily commuter trips, or local urban transit services. They even include a diversity of vehicles, operating speeds, right of way requirements, and service frequency. Passenger trains usually can be divided into two operations: intercity railway and intracity transit. Whereas as intercity railway involve higher speeds, longer routes, and lower frequency

Interior view of the top deck of a VR InterCity2 double-deck carriage

(usually scheduled), intracity transit involves lower speeds, shorter routes, and higher frequency (especially during peak hours).[30]

Intercity trains are long-haul trains that operate with few stops between cities. Trains typically have amenities such as a dining car. Some lines also provide over-night services with sleeping cars. Some long-haul trains been given a specific name. Regional trains are medium distance trains that connect cities with outlying, surrounding areas, or provide a regional service, making more stops and having lower speeds. Commuter trains serve suburbs of urban areas, providing a daily commuting service. Airport rail links provide quick access from city centres to airports.

High-speed rail are special inter-city trains that operates at much higher speeds than conventional railways, the limit being regarded at 200 to 320 km/h. High-speed trains are used mostly for long-haul service and most systems are in Western Europe and East Asia. The speed record is 574.8 km/h (357.2 mph), set by a modified French TGV.[31] [32] Magnetic levitation trains such as the Shanghai airport train use under-riding magnets which attract themselves upward towards the underside of a guideway and this line has achieved somewhat higher peak speeds in day-to-day operation than conventional high-speed railways, although only over short distances. Due to their heightened speeds, route alignments for high-speed rail tend to be steeper grades and broader curves compared to conventional railways. Their high kinetic energy translates to higher horsepower-to-ton ratios (20 hp/ton); this allows trains to accelerate and maintain higher speeds and negotiate steep grades as momentum builds up and recovered in downgrades (reducing cut, fill, and tunneling requirements). Since lateral forces act on curves, curvatures are designed with the highest possible radius. All these features are dramatically different from freight operations, thus justifying exclusive high-speed rail lines if it is economically feasible.[30]

Rapid transit is an intracity system built in large cities and has the highest capacity of any passenger transport system. It is grade separated and commonly built underground or elevated. At street level, smaller trams can be used. Light rails are upgraded trams that have step-free access, their own right-of-way and sometimes sections underground. Monorail systems operate as elevated, medium capacity systems. A people mover is a driverless, grade-separated train that serves only a few stations, as a shuttle. Due to the wide variety of rapid transit systems without much uniformity, route alignment vary widely with diverse right-of-ways (private land, side of road, street median) and geometric characteristics (sharp or broad curves, steep or gentle grades). For instance, the Chicago El trains are designed with extremely short cars to negotiate the sharp curves in the Loop. NJ's PATH have similar-sized cars to accommodate curves in the trans-Hudson tunnels. San Francisco's BART operate large cars on its well-engineered routes.[30]

Freight train

A freight train hauls cargo using freight cars specialized for the type of goods. Freight trains are very efficient, with economy of scale and high energy efficiency. However, their use can be reduced by lack of flexibility, if there is need of transshipment at both ends of the trip due to lack of tracks to the points of pick-up and delivery. Authorities often encourage the use of cargo rail transport due to its environmental profile.

Container trains have become the dominant type in the US for non-bulk haulage. Containers can easily be transshipped to other modes, such as ships and trucks, using cranes. This has succeeded the

Bulk cargo of minerals

boxcar (wagon-load), where the cargo had to be loaded and unloaded into the train manually. The intermodal containerization of cargo has revolutionized the supply chain logistics industry, reducing ship costs significantly. In Europe, the sliding wall wagon has largely superseded the ordinary covered wagons. Other types of cars include refrigerator cars, stock cars for livestock and autoracks for road vehicles. When rail is combined with road transport, a roadrailer will allow trailers to be driven onto the train, allowing for easy transition between road and rail.

Bulk handling represents a key advantage for rail transport. Low or even zero transshipment costs combined with energy efficiency and low inventory costs allow trains to handle bulk much cheaper than by road. Typical bulk cargo includes coal, ore, grains and liquids. Bulk is transported in open-topped cars, hopper cars and tank cars.

Infrastructure

Right of way

Railway tracks are laid upon land owned or leased by the railway company. Owing to the desirability of maintaining modest grades, rails will often be laid in circuitous routes in hilly or mountainous terrain. Route length and grade requirements can be reduced by the use of alternating cuttings, bridges and tunnels—all of which can greatly increase the capital expenditures required to develop a right of way, while significantly reducing operating costs and allowing higher speeds on longer radius curves. In densely urbanized areas, railways are sometimes laid in tunnels to minimize the effects on existing properties.

Trackage

Track consists of two parallel steel rails, anchored perpendicular to members called ties (sleepers) of timber, concrete, steel, or plastic to maintain a consistent distance apart, or rail gauge. Rail gauges are usually categorised as Standard gauge (1435 mm (4 ft 8 $\frac{1}{2}$ in)) used on approximately 60% of the world's existing railway lines, Broad gauge and Narrow gauge. In addition to the rail gauge, the tracks will be laid to conform with a Loading gauge which defines the maximum height and width for railway vehicles and their loads to ensure safe passage through bridges, tunnels and other structures.

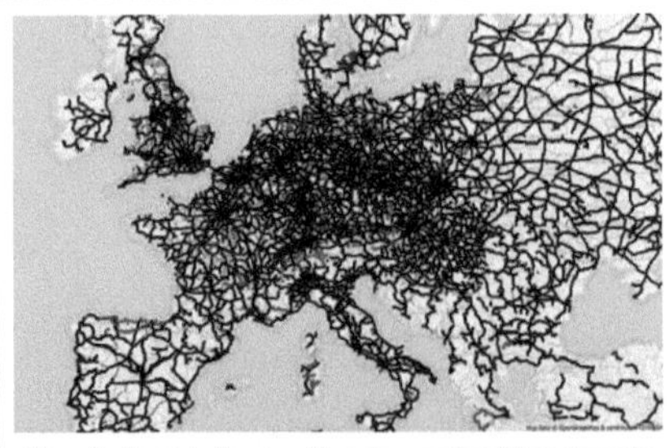

Map of railways in Europe with main operational lines shown in black, heritage railway lines in green and former routes in light blue

The track guides the conical, flanged wheels, keeping the cars on the track without active steering and therefore allowing trains to be much longer than road vehicles. The rails and ties are usually placed on a foundation made of compressed earth on top of which is placed a bed of ballast to distribute the load from the ties and to prevent the track from buckling as the ground settles over time under the weight of the vehicles passing above.

The ballast also serves as a means of drainage. Some more modern track in special areas is attached by direct fixation without ballast. Track may be prefabricated or assembled in place. By welding rails together to form lengths of continuous welded rail, additional wear and tear on rolling stock caused by the small surface gap at the joints between rails can be counteracted; this also makes for a quieter ride (passenger trains).

On curves the outer rail may be at a higher level than the inner rail. This is called superelevation or cant. This reduces the forces tending to displace the track and makes for a more comfortable ride for standing livestock and standing or seated passengers. A given amount of superelevation will be the most effective over a limited range of speeds.

Turnouts, also known as points and switches, are the means of directing a train onto a diverging section of track. Laid similar to normal track, a point typically consists of a frog (common crossing), check rails and two switch rails. The switch rails may be moved left or right, under the control of the signalling system, to determine which path the train will follow.

Spikes in wooden ties can loosen over time, but split and rotten ties may be individually replaced with new wooden ties or concrete substitutes. Concrete ties can also develop cracks or splits, and can also be replaced individually. Should the rails settle due to soil subsidence, they can be lifted by specialized machinery and additional ballast tamped under the ties to level the rails.

Long freight train crossing the Stoney Creek viaduct on the Canadian Pacific Railway in southern British Columbia

Periodically, ballast must be removed and replaced with clean ballast to ensure adequate drainage. Culverts and other passages for water must be kept clear lest water is impounded by the trackbed, causing landslips. Where trackbeds are placed along rivers, additional protection is usually placed to prevent streambank erosion during times of high water. Bridges require inspection and maintenance, since they are subject to large surges of stress in a short period of time when a heavy train crosses.

Train Inspection Systems

The inspection of railway equipment is essential for the safe movement of trains. Many types of defect detectors are in use on the world's railroads. These devices utilize technologies vary from a simplistic paddle and switch to infrared and laser scanning, and even ultrasonic audio analysis. Their use has avoided many rail accidents over the 70 years they have been used.

Signalling

Railway signalling is a system used to control railway traffic safely to prevent trains from colliding. Being guided by fixed rails with low friction, trains are uniquely susceptible to collision since

A Hot bearing detector w/ dragging equipment unit

they frequently operate at speeds that do not enable them to stop quickly or within the driver's sighting distance.

Most forms of train control involve movement authority being passed from those responsible for each section of a rail network to the train crew. Not all methods require the use of signals, and some systems are specific to single track railways.

The signalling process is traditionally carried out in a signal box, a small building that houses the lever frame required for the signalman to operate switches and signal equipment. These are placed at various intervals along the route of a railway, controlling specified sections of track. More recent technological developments have made such operational doctrine superfluous, with the centralization of signalling operations to regional control rooms. This has been facilitated by the increased use of computers, allowing vast sections of track to be monitored from a single location. The common method of block signalling divides the track into zones guarded by combinations of block signals, operating rules, and automatic-control devices so that only one train may be in a block at any time.

Electrification

The electrification system provides electrical energy to the trains, so they can operate without a prime mover onboard. This allows lower operating costs, but requires large capital investments along the lines. Mainline and tram systems normally have overhead wires, which hang from poles along the line. Grade-separated rapid transit sometimes use a ground third rail.

Power may be fed as direct or alternating current. The most common currencies are 600 and 750 V for tram and rapid transit systems, and 1,500 and 3,000 V for mainlines. The two dominant AC systems are 15 kV AC and 25 kV AC.

Stations

A railway station serves as an area where passengers can board and alight from trains. A goods station is a yard which is exclusively used for loading and unloading cargo. Large passenger stations have at least one building providing conveniences for passengers, such as purchasing tickets and food. Smaller stations typically only consist of a platform. Early stations were sometimes built with both passenger and goods facilities.[33]

Platforms are used to allow easy access to the trains, and are connected to each other via underpasses, footbridge and level crossings. Some large stations are built as cul-de-sac, with trains only operating out from one direction. Smaller stations normally serve local residential areas, and may have connection to feeder bus services. Large stations, in particular central stations, serve as the main public transport hub for the city, and have transfer available between rail services, and to rapid transit, tram or bus services.

Records

* The total number of Indian Railways stations is thought to be near about 8000.

Operations

Ownership

Traditionally, the infrastructure and rolling stock are owned and operated by the same company. This has often been by a national railway, while other companies have had private railways. Since the 1980s, there has been an increasing tendency to split up railway companies, with separate companies owning the stock from those owning the infrastructure, particularly in Europe, where this is required by the European Union. This has allowed open access by any train operator to any portion of the European railway network.

In the United States, railways, such as Union Pacific, are privately owned

In the U.S., virtually all rail networks and infrastructure are privately-owned with passenger lines, primarily Amtrak, operating as tenants on the freight lines. Consequently, operations must be closely synchronized and coordinated between freight and passenger railways for timeliness of their schedules. Due to this shared system, both are regulated by the Federal Railroad Administration (FRA) and follow the AREMA standards for track work and AAR standards for vehicles.[30]

Financing

The main source of income for railway companies is from ticket revenue (for passenger transport) and shipment fees for cargo. Discounts and monthly passes are sometimes available for frequent travellers. Freight revenue may be sold per container slot or for a whole train. Sometimes, the shipper owns the cars and only rents the haulage. For passenger transport, advertisement income can be significant.

Government may choose to give subsidies to rail operation, since rail transport has fewer externalities than other dominant modes of transport. If the railway company is state-owned, the state may simply provide direct subsidies in exchange for an increased production. If operations have been privatized, several options are available. Some countries have a system where the infrastructure is owned by a government agency or company—with open access to the tracks for any company that meets safety requirements. In such cases, the state may choose to provide the tracks free of charge, or for a fee that does not cover all costs. This is seen as analogous to the government providing free access to roads. For passenger operations, a direct subsidy may be paid to a public-owned operator, or public service obligation tender may be helt, and a time-limited contract awarded to the lowest bidder.

U.S.'s national passenger rail service, Amtrak, is a private railroad company chartered by the government. Similarly, Canada's VIA Rail system operates in the same fashion. As private passenger services lost significant ground in competition to the automobile and airplane and was forced out the market, they became stockholders of Amtrak either with a cash entrance fee or relinquishing their locomotives and rolling stock. Government aid supports Amtrak by supplying start-up capital and makes up for loses at end of the fiscal year.[29]

Safety

Rail transport is one of the safest forms of land travel.[34] Trains can travel at very high speed, but they are heavy, are unable to deviate from the track and require a great distance to stop. Possible accidents include derailment (jumping the track), a collision with another train or collision with an automobile or other vehicle at level crossings. The latter accounts for the majority of rail accidents and casualties. The most important safety measures to prevent accidents are strict operating rules, e.g. railway signalling and gates or grade separation at crossings. Train whistles, bells or horns warn of the presence of a train, while trackside signals maintain the distances between trains.

An important element in the safety of many high-speed inter-city networks such as Japan's Shinkansen is the fact that trains only run on dedicated railway lines, without level crossings. This effectively eliminates the potential for collision with automobiles, other vehicles and pedestrians, vastly reduces the likelihood of collision with other trains and helps ensure services remain timely.

Train crash at Montparnasse Station, Paris, France, in 1895

Maintenance

As in any infrastructure asset, railways must keep up with periodic inspection and maintenance in order to minimize effect of infrastructure failures that can disrupt freight revenue operations and passenger services. Because passengers are considered the most *crucial cargo* and usually operate at higher speeds, steeper grades, and higher capacity/frequency, their lines are especially important. Inspection practices include track geometry cars or walking inspection. Curve maintenance especially for transit services includes gauging, fastener tightening, and rail replacement. Rail corrugation is a common issue with transit systems due to the high number of light-axle, wheel passages which result in grinding of the wheel/rail interface. Since maintenance may overlap with operations, maintenance windows (nighttime hours, off-peak hours, altering train schedules or routes) must be closely followed. In addition, passenger safety during maintenance work (inter-track fencing, proper storage of materials, track work notices, hazards of equipment near states) must be regarded at all times. At times, maintenance access problems can emerge due to tunnels, elevated structures, and congested cityscapes. Here, specialized equipment, smaller versions of conventional maintenance gear are used.[30]

Unlike highways or road networks where capacity is disaggregated into unlinked trips over individual route segments, railway capacity is fundamentally considered a network system. As a result, many components are causes and effects of system disruptions. Maintenance must acknowledge the vast array of a route's performance (type of train service, origination/destination, seasonal impacts), line's capacity (length, terrain, number of tracks, types of train control), train's throughput (max speeds, acceleration/deceleration rates), and service features with shared passenger-freight tracks (sidings, terminal capacities, switching routes, and design type).[30]

Impact

Energy

Rail transport is an energy-efficient [35] but capital-intensive, means of mechanized land transport. The tracks provide smooth and hard surfaces on which the wheels of the train can roll with a minimum of friction. Moving a vehicle on and/or through a medium (land, sea, or air) requires overcoming resistance to motion. A land vehicle's total resistance (in pounds or Newtons) is a quadratic function of the vehicle's speed:

where:

BNSF Railway freight service in the United States

> R denotes total resistance
>
> a denotes initial constant resistance
>
> v denotes speed
>
> c denotes shape, frontal area, and sides of vehicle
>
> v^2 denotes effect of air resistance [30]

Essentially, resistance differs between vehicle's contact point and surface of roadway. Metal wheels on metal rails have a significant advantage of overcoming resistance compared to rubber-tired wheels on any road surface (railway - 0.001g at 10 mph and 0.024g at 60 mph; truck - 0.009g at 10 mph and 0.090 at 60 mph). In terms of cargo capacity combining speed and size being moved in a day (ton-miles/day):

German InterCityExpress

- human - can carry 100 lbs. 20 miles per day (one ton-mile/day)
- horse and wheelbarrow - can carry 4 ton-miles/day
- horse cart on good pavement - can carry 10 ton-miles/day
- fully utility truck - can carry 20,000 ton-miles/day
- long-haul train - can carry 500,000 ton-miles/day [30]

In terms of motive power, the horsepower and weight ratio, used to overcome resistance to motion when locomotives convert fuel to heat for propulsion a slow-moving barge requires 0.2 hp/net ton, a railway and pipeline requires 2.5 hp/net ton, and truck requires 10 hp/net ton. However, at higher speeds, a railway overcomes the truck and proves most economical. [30]

As an example, a typical modern wagon can hold up to 113 tonnes of freight on two four-wheel bogies. The contact area between each wheel and the rail is a strip no more than a few millimetres wide, which minimizes friction. The track distributes the weight of the train evenly, allowing significantly greater loads per axle and wheel than in road transport, leading to less wear and tear on the permanent way. This can save energy compared with other forms of transport, such as road transport, which depends on the friction between rubber tires and the road. Trains have a small frontal area in relation to the load they are carrying, which reduces air resistance and thus energy usage.

In addition, the presence of track guiding the wheels allows for very long trains to be pulled by one or a few engines and driven by a single operator, even around curves, which allows for economies of scale in both manpower and energy use; by contrast, in road transport, more than two articulations causes fishtailing and makes the vehicle unsafe.

Usage

Due to these benefits, rail transport is a major form of passenger and freight transport in many countries. In India, China, South Korea and Japan, many millions use trains as regular transport. It is widespread in European countries. Freight rail transport is widespread and heavily used in North America, but intercity passenger rail transport on that continent is relatively scarce outside the Northeast Corridor due to the loss of competition to other preferred modes, particular automobiles and airplanes.[29] [36]

Railway tracks running through Stanhope, United Kingdom

Africa and South America have some extensive networks such as in South Africa, Northern Africa and Argentina; but some railways on these continents are isolated lines. Australia has a generally sparse network befitting its population density, but has some areas with significant networks, especially in the southeast. In addition to the previously existing east-west transcontinental line in Australia, a line from north to south has been constructed. The highest railway in the world is the line to Lhasa, in Tibet,[37] partly running over permafrost territory. The western Europe region has the highest railway density in the world, and has many individual trains which operate through several countries despite technical and organizational differences in each national network.

Social/Economic Benefits

Railways have also been shown to contribute to social vibrancy and economic competitiveness in its ability to transport large amounts of customers and workers to city centers and inner suburbs (i.e. Washington DC as a cultural/policy center due to metrorail system, San Francisco's lively downtown due to the BART system). Hong Kong has recognized rail as "the backbone of the public transit system" and as such developed their franchised bus system and road infrastructure in compherensive alignment with their rail services.[38] China's large cities including Beijing, Shanghai, and Guangzhou recognize rail transit lines as the framework and bus lines as the main body to their metropolitan transportation systems.[39] The Japanese Shinkansen was built to meet the growing traffic demand in the "heart of Japan's industry and economy" situated on the Tokyo-Kobe line.[40]

Japanese Shinkansen

As opposed to highway expansion, indicative of the U.S. transportation policy, that incentivizes development of suburbs at the periphery, contributing to increased vehicle miles traveled, carbon emissions, development of greenfield spaces, and depletion of natural reserves, railways channel growth toward dense city agglomerations and along its artery. These sustainable arrangements revalue city spaces, local taxes, housing values, and promotion of mixed use development.[41] [42]

See also

- Environmental design in rail transportation
- International Union of Railways
- List of rail transport topics
- List of railway companies
- Megaproject
- Passenger rail terminology
- Rail transport by country
- Railway systems engineering

References

Notes

[1] According to this source (http://www.railwatch.org.uk/backtrack/rw94/rw094p06.pdf), railways are safest on both a per-mile and per-hour basis, whereas air transport is safe only on a per-mile basis

[2] Lewis, M. J. T. "Railways in the Greek and Roman World" (http://www.sciencenews.gr/docs/diolkos.pdf) (pdf). . Retrieved 11 April 2009.

[3] Hylton, Stuart (2007). *The Grand Experiment: The Birth of the Railway Age 1820-1845*. Ian Allan Publishing.

[4] Kriechbaum, Reinhard (15 May 2004). "Die große Reise auf den Berg" (http://www.die-tagespost.de/Archiv/titel_anzeige.asp?ID=8916) (in German). *der Tagespost*. . Retrieved 22 April 2009.

[5] "Der Reiszug - Part 1 - Presentation" (http://www.funimag.com/funimag10/RESZUG01.HTM). Funimag. . Retrieved 22 April 2009.

[6] Georgius Agricola (1913). *De re metallica*. ISBN 0486600068.

[7] Porter, Peter (1914). *Landmarks of the Niagara Frontier*. The Author. ISBN 0665783477.

[8] Morlok, Edward K. (11 May 2005). "First permanent railroad in the U.S. and its connection to the University of Pennsylvania" (http://www.seas.upenn.edu/~morlok/morlokpage/transp_data.html). . Retrieved 19 September 2007.

[9] at Coalbrookdale *Railways (pt 1)* (http://www.1902encyclopedia.com/R/RAI/railway-01.html). Encyclopedia Britannica. 1902. ISBN 187252463X. . Retrieved 2011-02-15.

[10] Vaughan, A. (1997). *Railwaymen, Politics and Money*. London: John Murray. ISBN 0719557461.

[11] Marshall, John (1979). *The Guiness Book of Rail Facts & Feats*. ISBN 0-900424-56-7.

[12] "Surrey Iron Railway 200th - 26th July 2003" (http://www.stephensonloco.fsbusiness.co.uk/surreyiron.htm). *Early Railways*. Stephenson Locomotive Society. . Retrieved 19 September.

[13] Skempton, A.W. (2002). *A biographical dictionary of civil engineers in Great Britain and Ireland, John Birkinshaw* (http://books.google.com/books?id=jeOMfpYMOtYC&pg=PA59). pp. 59–60. ISBN 9780727729392. .

[14] "The History of the Tom Thumb - Peter Cooper" (http://inventors.about.com/library/inventors/bl_tom_thumb.htm). Inventors.about.com. . Retrieved 8 May 2011.

[15] Storey, Steve. "Charleston & Hamburg Railroad" (http://www.railga.com/charlhmbrg.html). Railga.com. . Retrieved 8 May 2011.

[16] "Richard Trevithick's steam locomotive" (http://www.museumwales.ac.uk/en/rhagor/article/trevithic_loco/). Museumwales.ac.uk. 15 December 2008. . Retrieved 8 May 2011.

[17] "Steam train anniversary begins" (http://news.bbc.co.uk/2/hi/uk_news/wales/3509961.stm). BBC. 21 February 2004. . Retrieved 8 May 2011. "A south Wales town has begun months of celebrations to mark the 200th anniversary of the invention of the steam locomotive. Merthyr Tydfil was the location where, on 21 February 1804, Richard Trevithick took the world into the railway age when he set one of his high-pressure steam engines on a local iron master's tram rails"

[18] Payton, Philip (2004). *Oxford Dictionary of National Biography*. Oxford University Press.

[19] Chartres, J.. "Richard Trevithick". In Cannon, John. *Oxford Companion to British History*. p. 932.

[20] "Early Days of Mumbles Railway" (http://www.bbc.co.uk/wales/southwest/sites/swansea/pages/mumbles_trainanniv.shtml). BBC. 15 February 2007. . Retrieved 19 September 2007.

[21] "John Blenkinsop" (http://www.britannica.com/EBchecked/topic/69274/John-Blenkinsop). *Encyclopædia Britannica*. . Retrieved 8 May 2011.

[22] Ellis, Hamilton (1968). *The Pictorial Encyclopedia of Railways*. Hamlyn Publishing Group.

[23] "Liverpool and Manchester" (http://www.spartacus.schoolnet.co.uk/RAliverpool.htm). . Retrieved 19 September 2007.

[24] Dilts, James D. (1996). *The Great Road: The Building of the Baltimore and Ohio, the Nation's First Railroad, 1828-1853* (http://books.google.com/?id=JjrCWPwvHzIC&lpg=PR18&dq=dilts b&o&pg=PA26#v=onepage&q=first). Palo Alto, CA: Stanford University Press. p. 26. ISBN 978-0804726290. .

[25] "The Journal of Ebenezer Mattoon Chamberlain 1832-5", *Indiana Magazine of History*, Vol. XV, September, 1919, No. 3, p.233ff.

[26] Wolmar (2009) p.xiii.

[27] Ambrose, Stephen E. (2000). *Nothing Like It In The World; The men who built the Transcontinental Railroad 1863-1869* (http://books. google.com/?id=TZp_GT7PscIC&lpg=PP1&dq=ambrose nothing like it&pg=PP1#v=onepage&q=). Simon & Schuster. ISBN 0-684-84609-8. .

[28] Tokle, Bjørn (2003) (in Norwegian). *Communication gjennom 100 år*. Meldal: Chr. Salvesen & Chr. Thams's Communications Aktieselskab. p. 54.

[29] EuDaly, K, Schafer, M, Boyd, Jim, Jessup, S, McBridge, A, Glischinksi, S. (2009). The Complete Book of North American Railroading. Voyageur Press. 1-352 pgs.

[30] American Railway Engineering and Maintenance of Way Association Committee 24 - Education and Training. (2003). Practical Guide to Railway Engineering. AREMA, 2nd Ed.

[31] Associated Press (4 April 2007). "French train breaks speed record" (http://web.archive.org/web/20070407194558/http://www.cnn. com/2007/WORLD/europe/04/03/TGVspeedrecord.ap/index.html). *CNN*. Archived from the original (http://www.cnn.com/2007/ WORLD/europe/04/03/TGVspeedrecord.ap/index.html) on 7 April 2007. . Retrieved 3 April 2007.

[32] Fouquet, Helene and Viscousi, Gregory (3 April 2007). "French TGV Sets Record, Reaching 357 Miles an Hour (Update2)" (http://www. bloomberg.com/apps/news?pid=newsarchive&sid=aW23Aw20niIo&refer=europe). Bloomberg L.P.. . Retrieved 8 May 2011.

[33] "The Inception of the English Railway Station". *Architectural History* (SAHGB Publications Limited) **4**: 63–76. 1961. doi:10.2307/1568245. JSTOR 1568245.

[34] U.S. Bureau of Transportation Statistics (2010). *National Transportation Statistics. Table 2-1: Transportation Fatalities by Mode* (http:// www.bts.gov/publications/national_transportation_statistics/html/table_02_01.html) (Report). . Retrieved 2010-02-14.

[35] American Association of Railroads. "Railroad Fuel Efficiency Sets New Record" (http://www.progressiverailroading.com/news/article. asp?id=16740). . Retrieved 12 April 2009.

[36] "Public Transportation Ridership Statistics" (http://web.archive.org/web/20070815101950/http://www.apta.com/research/stats/ ridership/). American Public Transportation Association. 2007. Archived from the original (http://www.apta.com/research/stats/ridership/) on 15 August 2007. . Retrieved 10 September 2007.

[37] "New height of world's railway born in Tibet" (http://news.xinhuanet.com/english/2005-08/24/content_3397297.htm). Xinhua News Agency. 24 August 2005. . Retrieved 8 May 2011.

[38] Hong Kong Information Services Department of the Hong Kong SAR Government. Hong Kong 2009

[39] Hau H., Yun-feng G., Zhi-gang, L., Xiao-guang, Y. (2010). Effect of Integrated Multi-Modal Transit Information on Modal Shift. Intelligent Transportation Systems (ITSC), 2010 13th International IEEE Conference. 1753-1757pg.

[40] Nishida, M., The Shinkansen High-Speed Rail Network of Japan. Proceedings of an IIASA Conference, June 27–30, 1977.

[41] Squires, G. Ed. (2002) Urban Sprawl: Causes, Consequences, & Policy Responses. The Urban Institute Press.

[42] Puentes, R. (2008). A Bridge to Somewhere: Rethinking American Transportation for the 21st Century. Brookings Institution Metropolitan Policy Report: Blueprint for American Prosperity series report.

References

pfl:Aisebahn

Jesús Carranza, Veracruz

Jesus Carranza	
— Town & Municipality —	
Coordinates: 17°26′07″N 95°01′58″W	
Country	▌•▌ Mexico
State	State of Veracruz
Founded	
Municipal Status	1879
Government	
- Municipal President	Leobardo Benítez Osorio (1998-2000)
Elevationof seat	24 m (79 ft)
Population (2000)Municipality	
- Municipality	25,513
Time zone	CST (UTC-6)
Postal code (of seat)	96950
Demonym	

Jesus Carranza is a town and municipality located in the State of Veracruz, in Mexico. The town has an area of 486.32 kilometers squared. It represents about 0.67 percent of the State of Veracruz. Jesus Carranza borders the State of Oaxaca to its west. It is located 435 kilometers from the state capital of Veracruz, Xalapa. The municipality of Jesus Carranza governs the communities of Jesus Carranza, Suchilapan River, Col. Nuevo Morelos, El Tepache and Coapiloloyita.

History

Jesus Carranza began as a municipality with the name Suchilapam on December 4, 1879. On October 13, 1910 the town was named Santa Lucrecia by the religious community in the area. On November 5, 1932 the town and municipality was given its current name in honor of the revolutionary general, Jesus Carranza. Jesus Carranza was known for helping Benito Juarez during the Mexican revolution. One of his sons, Venustiano Carranza later became President of Mexico.

Olmec

Jesus Carranza is one of the areas associated with the ancient Pre-Columbian people, the Olmec. A famous Olmec statue, Senor de las Limas, was discovered in this area in 1965.

Land

Climate

Jesus Carranza is located in the foothills of the Sierra Madre Oriental and has an average temperature of 27 degrees Celsius.

Animals

There are many small animals that exist in the area such as wild boar, deer, iguanas, rabbit, reptiles, birds and armadillos.

Resources

Some of the natural resources in the area consist of sand, gravel, mahogany, cedar, zapote, ceiba. There are many marine resources such as snook, turtles and shrimp.

Agriculture

Most of the land in this area, 90 percent, is used for agriculture. The other ten percent is dedicated to occupied housing.

People

The population in Jesus Carranza, according to the 2000 Census, was 25,513 people. The majority of the inhabitants in this area are Spanish speaking. Within the total population exists about 2268 indigenous people. The main language among this group is Chinanteco.

Religion

There are approximately 17,000 people whose main religion is Catholisism. There are also about 2700 Protestants, 700 that practice other religions and around 2600 people who do not practice any of the recognized religions.

Current Events

On October 28, 2007 a Pemex oil pipeline that passes through Jesus Carranza ruptured. Pemex is Mexico's state owed petroleum company. It was reported that 10,000 barrels of oil leaked into the states three main waterways. This contamination caused significant environmental impacts in the area.

References

- "Translated version of http://www.e-mexico.gob.mx/work/EMM04/Veracruz/mpios/30091a.htm." Google Translate. 6 Feb. 2009
  <http://translate.google.com/translate?hl=en&sl=es&u=http://www.e-mexico.gob.mx/work/EMM04/Veracruz/mpios/30091a.h>
- "Photo from Reuters Pictures - Daylife." Daylife - A New Way to Explore the World. 8 Feb. 2009
  <http://www.daylife.com/photo/01z7gcK8PC7lk>.

External links

- Official page of the municipality of Jesús Carranza [1]
- (Spanish) Municipal Official Information [2]

References

[1] http://www.jesuscarranza.gob.mx/
[2] http://portal.veracruz.gob.mx/portal/page?_pageid=153,4493464&_dad=portal&_schema=PORTAL&municipio=jesuscarranza.pdf

Politics of Mexico

The **politics of Mexico** take place in a framework of a federal presidential representative democratic republic whose government is based on a congressional system, whereby the president of Mexico is both head of state and head of government, and of a multi-party system. The federal government represents the United Mexican States and is divided into three branches: executive, legislative and judicial, as established by the Political Constitution of the United Mexican States, published in 1817. The constituent states of the federation must also have a republican form of government based on a congressional system as established by their respective constitutions.

The executive power is exercised by the executive branch, which is headed by the President, advised by a cabinet of secretaries that are independent of the legislature. Legislative power is vested upon the Congress of the Union, a two-chamber legislature comprising the Senate and the Chamber of Deputies. Judicial power is exercised by the judiciary, consisting of the Supreme Court of Justice of the Nation, the Council of the Federal Judiciary and the collegiate, unitary and district tribunals.

The politics of Mexico are dominated by three political parties: National Action Party (PAN), the Party of the Democratic Revolution (PRD) and Institutional Revolutionary Party (PRI).

Political parties

Constitutionally, political parties in Mexico must promote the participation of the people in the democratic life of the country, contribute in the representation of the nation and citizens, and be the access through which citizens can participate in public office, through whatever programs, principles and ideals they postulate.[1] All political parties must be registered before the Federal Electoral Institute (IFE), the institution in charge of organizing and overseeing the federal electoral processes, but must obtain at least 2% of votes in the federal elections to keep their registry. Registered political parties receive public funding for their operation and can also obtain private funding within the limits prescribed by the law. As of 2010 the following political parties are registered before the IFE and all have representatives at the Congress of the Union:

- Institutional Revolutionary Party (*Partido Revolucionario Institucional*, PRI), founded in 1929;
- National Action Party (*Partido Acción Nacional*, PAN), founded in 1939;
- Party of the Democratic Revolution (*Partido de la Revolución Democrática*, PRD), founded in 1989;
- Labor Party (*Partido del Trabajo*, PT), founded in 1990;
- Green Ecological Party (*Partido Verde Ecologista de México*, PVEM), founded in 1986, but lost its registry on two consecutive elections; it has retained its registry since 1993;
- Convergence Party (*Convergencia*), founded in 1997;
- New Alliance (*Nueva Alianza*, PNA or PANAL), founded in 2005;

Political parties are allowed to form alliances or coalitions to nominate candidates for any particular election. The coalition must present itself with a particular name and logo. Proportional representation (plurinominal) seats are assigned to the coalition based on the percentage of votes obtained in the elections, and then the coalition re-assigns them to the constituent political parties. Once each party in the coalition has been assigned plurinominal seats, they do not necessarily continue to work as a coalition in government.

Throughout the 20th century, PRI had an almost hegemonic power at the state and federal level, which slowly began to recede in the late 1980s. Even though since the 1940s, PAN had won a couple of seats in the Congress, and in 1947 the first presidential municipality (in Quiroga, Michoacán),[2] it wasn't until 1989, that the first non-PRI governor of a state was elected (at Baja California). It was in 1997, that PRI lost its absolute majority at the Congress of the Union, and in 2000 the first non-PRI president was elected since 1929.

Elections and political composition of the institutions

Archivo General de la Nacion

Suffrage is universal, free, secret and direct for all Mexican citizens 18 and older, and is compulsory (but not enforced). The identity document in Mexico serves also as the voting card, so all citizens are automatically registered for all elections; that is, no pre-registration is necessary for every election. All elections are direct; that is, no electoral college is constituted for any of the elections at the federal, state or municipal level. Only when an incumbent president is absolutely absent (either through resignation, impeachment or death), the Congress of the Union constitutes itself acts as an electoral college to elect an interim president by absolute majority.

Presidential elections are scheduled every six years, except in the exceptional case of absolute absence of the president. Legislative elections are scheduled every six years for the Senate, to be fully renewed in elections held concurrently with the presidential elections; and every three years for the Chamber of Deputies. Elections are usually held on the first Sunday of July. State governors are also elected every six years, whereas the legislatures are renewed every three years. State elections need not be concurrent with federal elections. Federal elections are organized and supervised by the autonomous public Federal Electoral Institute, whereas state and municipal elections are organized and supervised by electoral institutes constituted by each state of the federation. Elections within the Federal District are also organized by a local electoral institute.

A strongly ingrained concept in Mexican political life is "no reelection." The theory was implemented after Porfirio Díaz managed to monopolize the presidency for over 25 years. Presently, Mexican presidents are limited to a single six-year term, and no one who has held the office even on a caretaker basis is allowed to hold the office again. Deputies and senators are not allowed to immediately succeed themselves.

Federal elections

The most recent federal presidential elections were held on July 2, 2006 concurrent with the full renovation of both chambers of the Congress of the Union. In these elections the Party of the Democratic Revolution (PRD), the Labour Party (PT) and Convergence (CV) formed a coalition called Coalition for the Good of All. The Institutional Revolutionary Party (PRI) and the Ecologist Green Party (PVEM) formed a coalition called Alliance for Mexico.

Presidential elections

The presidential elections were the most competitive in the history of the country in which the difference in the ballot count between the winner and the first runner up was less than one percent point, and in which neither candidate got absolute majority in a system in which a second round of voting has not been instituted. Felipe Calderón got the greatest number of votes according to the preliminary computation (PREP) and the ballot recount. Andrés Manuel López Obrador contested the results and demanded a vote-per-vote recount, which was denied by the Federal Electoral Tribunal, based on the argument that inconsistencies could not be proved for all electoral circumscriptions, but order a partial recount of votes of those

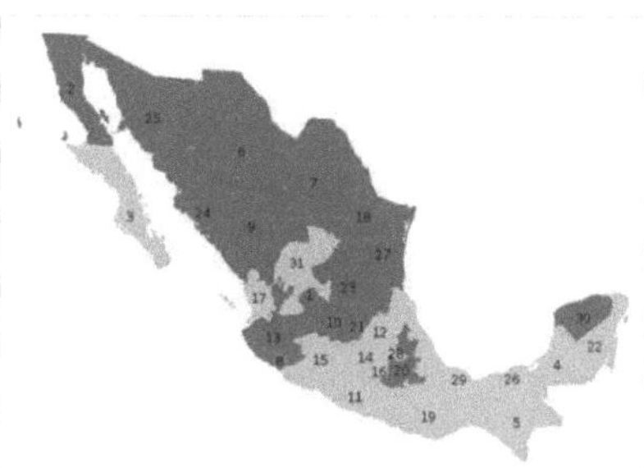

Majority candidate per state according to PREP. Blue: Felipe Calderón, Yellow: Andrés Manuel López Obrador

that did show inconsistencies which represented 9.2% of the total, after which the results were not significantly altered. The Federal Electoral Tribunal declared Felipe Calderón the winner of the elections on September 5, and president elect. He took office on December 1, and his term will end on November 30, 2012.

Summary of the 2 July 2006 Mexican presidential election results

Candidates	Party	Votes	%
Felipe Calderón	National Action Party	15,000,284	35.89%
Andrés Manuel López Obrador	Coalition for the Good of All (PRD, PT, CV)	14,756,350	35.31%
Roberto Madrazo	Alliance for Mexico (PRI, PVEM)	9,301,441	22.26%
Patricia Mercado	Social Democratic and Peasant Alternative Party	1,128,850	2.70%
Roberto Campa Cifrián	New Alliance	401,804	0.96%
Write in		297,989	0.71%
Blank/Invalid		904,604	2.16%
Total		**41,791,322**	**100.0%**
Source: Instituto Federal Electoral [3]			

Congressional elections

The concurrent congressional elections were not contested by any party. Both chambers were completely renewed and no party obtained absolute majority. All deputies and senators took office on September 1. First-past-the-post plurality candidates (FPP) of coalitions represent the parties of which they are members. Proportional representation (PR) seats assigned to coalitions were further reassigned to their constituent parties in whatever manner and number they agreed upon. Parties that formed a coalition for the general elections may continue to work together but they do not form a unified political bloc at the Congress; parliamentary groups are identified by parties and not by coalitions.[4]

Summary of the 2 July 2006 Mexican Chamber of Deputies election results

Parties and/or coalitions		Votes	%	FPP	PR	Total seats
National Action Party (*Partido Acción Nacional*)		13,845,121	33.41	137	69	206
Coalition for the Good of All (*Coalición por el Bien de Todos*)	Party of the Democratic Revolution (*Partido de la Revolución Democrática*)	12,013,364	28.99	91	36	127
	Convergence (*Convergencia*)			5	12	17
	Labour Party (*Partido del Trabajo*)			2	10	12
	No party			0	1	1
Alliance for Mexico (*Alianza por México*)	Institutional Revolutionary Party (*Partido Revolucionario Institucional*)	11,676,585	28.18	65	41	106
	Ecologist Green Party of Mexico (*Partido Verde Ecologista de México*)			0	17	17
New Alliance Party (*Partido Nueva Alianza*)		1,883,476	4.55	0	9	9
Social Democratic and Peasant Alternative Party (*Partido Alternativa Socialdemócrata y Campesina*)		850,989	2.05	0	4	4
Total		41,435,934	100.00	300	200	**500**

Source: Chamber of Deputies [5]

The 64 Senate first-past-the-post (FPP) seats are assigned to the pair of senators of the same party (who run together) that obtain the majority of votes per state and the Federal District. The 32 first minority (FM) seats are assigned to the first runner-up per party and the Federal District. Finally, 32 proportional representation (PR) seats are assigned according to national votes to the party or coalition in relation to the total number of votes obtained nationally. PR seats are assigned to the coalition who then reassigns them to its constituent parties in whatever manner and number they had originally agreed upon, and may or may not work as a bloc in the Senate.

Summary of the 2 July 2006 Mexican Senate election results

Parties and/or coalitions		Votes	%	FPP	FM	PR	Total seats
National Action Party (*Partido Acción Nacional*)		14,035,503	33.63	32	9	11	52
Coalition for the Good of All (*Coalición por el Bien de Todos*)	Party of the Democratic Revolution (*Partido de la Revolución Democrática*)	12,397,008	29.70	22	4	5	31
	Labour Party (*Partido del Trabajo*)			0	0	3	3
	Convergence (*Convergencia*)			0	0	2	2
Alliance for Mexico (*Alianza por México*)	Institutional Revolutionary Party (*Partido Revolucionario Institucional*)	11,681,395	27.99	10	19	6	35
	Ecologist Green Party of Mexico (*Partido Verde Ecologista de México*)			0	0	4	4
New Alliance Party (*Partido Nuevo Alianza*)		1,688,198	4.04	0	0	1	1

Social Democratic and Peasant Alternative Party (*Partido Alternativa Socialdemócrata y Campesina*)	795,730	1.91	0	0	0	0
Total	41,739,188	100.00	64	32	32	**128**

Source: Senate [6]

State elections

The elections in each state are done at different times, depending on the state, and are not necessarily held at the same time with the federal elections. Currently, even though the PRI is the third political force in the Congress of the Union, in terms of number of seats, it is still the first political force in terms of the number of states governed by it. As of 2010:

States and political party in power

- PRI governs 20 states: Aguascalientes, Campeche, Chihuahua, Coahuila, Colima, Durango, Hidalgo, México, Nayarit, Nuevo León, Puebla, Querétaro, Quintana Roo, San Luis Potosí, Sinaloa, Tabasco, Tamaulipas, Veracruz, Yucatán and Zacatecas.
- PAN governs 7 estates: Baja California, Baja California sur, Guanajuato, Jalisco, Morelos, Sonora and Tlaxcala.
- PRD governs 3 states and the Federal District (Mexico City): Chiapas, Distrito Federal, Guerrero and Michoacán.
- Convergencia governs a state: Oaxaca.

Historical political development

Vicente Fox, president 2000-2006

In 1929, all factions and generals of the Mexican Revolution were united into a single party, the National Revolutionary Party (NRP), with the aim of stabilizing the country and ending internal conflicts. During the following administrations, since 1928, many of the revolutionary ideals were put into effect, among them the free distribution of land to peasants and farmers, the nationalization of the oil companies, the birth and rapid growth of the Social Security Institute as well as that of Labor Unions, and the protection of national industries. The party was later renamed the Mexican Revolution Party and finally the Institutional Revolutionary Party. The social institutions created by the party itself provided it with the necessary strength to stay in power. In time, the system gradually became, as some political scientists have labeled it, an "electoral authoritarianism",[7] in that the party resorted to any means necessary, except that of the dissolution of the constitutional and electoral system itself, to remain in power. In fact, Mexico was considered a bastion of continued constitutional government in times where coup d'états and military dictatorships were the norm in Latin America, in that the institutions were renovated electorally, even if only in appearance and with little participation of the opposition parties at the local level.

The first cracks in the system, even though they were merely symbolic, were the 1970s reforms to the electoral system and the composition of the Congress Pof the Union which for the first time incorporated proportional representation seats allowing opposition parties to obtain seats, though limited in number, in the Chamber of

Deputies. As minority parties became involved in the system, they gradually demanded more changes, and a full democratic representation. Even though in the 1960s, a couple (of more than two thousand) municipalities were governed by opposition parties, the first state government to be won by an opposition party was Baja California, in 1989.

The presidential elections held in 1988 marked a watershed in Mexican politics, as they were the first serious threat to the party in power by an opposition candidate, Cuauhtémoc Cárdenas, a defector from the ruling Institutional Revolutionary Party (PRI) and son of former President Lazaro Cardenas, who was nominated by a broad coalition of leftist parties. He officially received 31.1 percent of the vote, against 50.4 percent for Carlos Salinas de Gortari, the PRI candidate, and 17 percent for Manuel Clouthier of the National Action Party (PAN). It was widely said that Cardenas had won the election, but that the then government-controlled electoral commission had altered the results after the infamous "glitch in the system" (*se cayó el sistema*, as it was reported). In the concurrent elections, the PRI came within 11 seats of losing the majority of Chamber of Deputies, and opposition parties captured 4 of the 64 Senate seats - the first time that the PRI had failed to hold every seat in the Senate. Capitalizing on the popularity of President Salinas, however, the PRI rebounded in the mid-term congressional elections of 1991, winning 320 seats.

Subsequent changes included the creation of the Federal Electoral Institute in the 1990s and the inclusion of proportional representation and first minority seats in the Senate. The presidential election of 1994 was judged to be the first relatively free election in modern Mexican history. Ernesto Zedillo of the PRI won with 50.2 percent of the vote, against 26.7 percent for Diego Fernández de Cevallos of PAN and 17.1 percent for Cardenas, who this time represented the Party of the Democratic Revolution (PRD). Although the opposition campaign was hurt by the desire of the Mexican electorate for stability, following the assassination of Luis Donaldo Colosio, the intended PRI candidate, and the recent outbreak of hostilities in the state of Chiapas, Zedillo's share of the vote was the lowest official percentage for any PRI presidential candidate up to that time.

In the 1997 mid-term elections, no party held majority in the Chamber of Deputies, and in 2000 the first opposition party president was sworn in office since 1929. Vicente Fox won the election with 43% of the vote, followed by PRI candidate Francisco Labastida with 36%, and Cuauhtémoc Cárdenas of the Party of the Democratic Revolution (PRD) with 17%.

Numerous electoral reforms implemented after 1989 aided in the opening of the Mexican political system, and opposition parties made historic gains in elections at all levels. Many of the current electoral concerns have shifted from outright fraud to campaign fairness issues. During 1995-96 the political parties negotiated constitutional amendments to address these issues. Implementing legislation included major points of consensus that had been worked out with the opposition parties. The thrust of the new laws has public financing predominate over private contributions to political parties, tightens procedures for auditing the political parties, and strengthens the authority and independence of electoral institutions. The court system also was given greatly expanded authority to hear civil rights cases on electoral matters brought by individuals or groups. In short, the extensive reform efforts have "leveled the playing field" for the parties.

The 2006 elections saw the PRI fall to third place behind both the PAN and the PRD. Roberto Madrazo, the presidential candidate, polled only 22.3 percent of the vote, and the party ended up with only 121 seats in the Chamber of Deputies, a loss of more than half of what the party had obtained in 2003, and 38 Senate seats, a loss of 22. Nevertheless, at the state level, more states are still governed by PRI than by the rest of the parties.

References and notes

[1] Article 41, Political Constitution of the United Mexican States (http://constitucion.gob.mx/index.php?art_id=41)

[2] Efemérides del PAN (http://www.pan.org.mx/?P=363)

[3] http://www.ife.org.mx/documentos/Estadisticas2006/presidente/nac.html

[4] Composición de Grupos Parlamentarios (http://www3.diputados.gob.mx/camara/001_diputados/005_grupos_parlamentarios)

[5] http://www3.diputados.gob.mx/camara/001_diputados/005_grupos_parlamentarios

[6] http://www.senado.gob.mx/legislatura.php?ver=listadox

[7] Using the phrase of Schedler A (2004) *From Electoral Authoritarianism to Democratic Consolidation" in* Mexico's Democracy at Work, *Crandall R, Paz G, Roett R (editors), Lyenne Reinner Publisher, Colorado USA*

See also

- State governments of Mexico
- Federal government of Mexico
- Powers of the Union (Mexico)

External links

- Presidency of the United Mexican States (http://www.presidencia.gob.mx)
- Congress of the Union (http://www.congreso.gob.mx)
- Supreme Court of Justice of the Nation (http://www.scjn.gob.mx)
- Mexican Council for Economic and Social Development (http://www.consejomexicano.com/)
- Mexico Development Gateway (http://www.portaldeldesarrollo.info/)

Ferrocarril Transistmico

<table>
<tr><td colspan="2" align="center">Ferrocarril Transistmico</td></tr>
<tr><td colspan="2">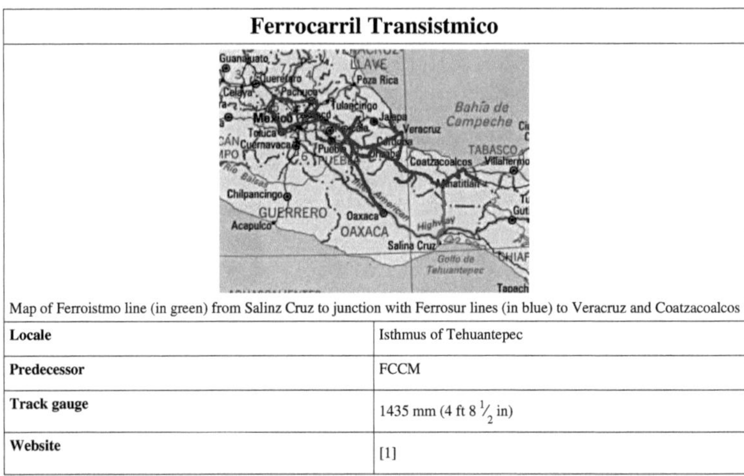
Map of Ferroistmo line (in green) from Salinz Cruz to junction with Ferrosur lines (in blue) to Veracruz and Coatzacoalcos</td></tr>
<tr><td>Locale</td><td>Isthmus of Tehuantepec</td></tr>
<tr><td>Predecessor</td><td>FCCM</td></tr>
<tr><td>Track gauge</td><td>1435 mm (4 ft 8 $\frac{1}{2}$ in)</td></tr>
<tr><td>Website</td><td>[1]</td></tr>
</table>

The **Ferrocarril Transistmico**, also known as **Ferrocarril del Istmo de Tehuantepec, S.A. de C.V.** or simply **Ferroistmo**,[2] today is a railroad with no rolling stock, owned by the Mexican government, that crosses the Isthmus of Tehuantepec between Puerto Mexico, Veracruz, and Salina Cruz, Oaxaca. It is leased to Ferrocarril del Sureste FERROSUR. It was formerly leased to Ferrocarriles Chiapas-Mayab until Genesee & Wyoming gave up its concession in 2007.[3] Originally it was known as the **Tehuantepec Railway**.

See also

- American-Hawaiian Steamship Company
- Ferrocarril de Veracruz al Istmo
- List of Mexican railroads

References

[1] http://www.ferroistmo.com.mx/

[2] McManus, Lowell G.. "The Mexican Railways" (http://www.mexlist.com/railways.htm). . Retrieved 2011-01-31.

[3] Central America going nowhere (http://www.railwaygazette.com/nc/news/single-view/view/central-america-going-nowhere.html). *Railway Gazette International* August 2007.

Ferrocarriles Nacionales de México

Ferrocarriles Nacionales de México, (better known as *N de M*) was Mexico's state owned railroad company from 1938 to 1998, and prior to 1938 (dating from the regime of Porfirio Díaz) a major railroad controlled by the government that linked Mexico City to the major cities of Nuevo Laredo and Ciudad Juárez on the U.S. border. The first trains to Nuevo Laredo from Mexico City began operating in 1903.[1]

N de M absorbed the Mexican Central Railroad (*Ferrocarril Central Mexicano*, first section from Mexico City to León, Guanajuato, opened in 1882) in 1909, thus acquiring a second border gateway at Ciudad Juárez (adjacent to El Paso, Texas). The N de M was nationalized by President Lázaro Cárdenas del Río in 1938, and privatized 60 years later by President Ernesto Zedillo. N de M operated most railway trackage through the central and northeastern regions of the republic.

During the days of steam locomotives, N de M was best known for operating *Niágara* class locomotives, which took their name from the New York Central Railroad locomotives of the same wheel configuration. N de M was one of the few railroads outside the US to purchase new diesel locomotives from Baldwin Locomotive Works, including the Baldwin DR-6 and Centipede.

Ferrocarriles Nacionales de Mexico train.

A train in Mexico City station, 1984

In Acámbaro, Guanajuato, N de M operated one of the few facilities in Latin America that was capable of constructing and doing complete rebuilds of steam locomotives, thus with rare exceptions (as with the *Niagaras*) , most of N de M steam motive power was purchased used and rebuilt there. Portions of the facility and a preserved 2-8-0 steam locomotive remain as part of Acambaro's municipal railway museum.

In 1995, the Mexican government announced that the FNM would be privatized and divided into four main systems, see Rail transport in Mexico: Privatization. As part of the restructuring for privatization, FNM suspended passenger rail service in 1997, and the new arrangements applied from 1998. The companies were Kansas City Southern de Mexico, Ferromex, Ferrosur, and (owned jointly by the three companies) Ferrocarril y Terminal del Valle de México or Ferrovalle which operates railroads and terminals in and around Mexico City.

Logo, Painting on a train

As of 2006, the remaining parts of NdeM are in the process of liquidation.[2]

Notable named passenger trains of the N de M

Named trains usually bore names related to the destination, for example, *El Purépecha* referred to the Purepecha (Tarascan) peoples of western Michoacán.

- *Águila Azteca* - Mexico City - Monterrey - Nuevo Laredo
- *El Regiomontano* - Mexico City - Monterrey - Nuevo Laredo
- *El Fronterizo* - Ciudad Juárez - Chihuahua - Mexico City
- *El Oaxaqueño* - Mexico City - Puebla - Oaxaca
- *El Purépecha* - Mexico City - Morelia - Uruapan (Michoacán)
- *El Tapatío* - Mexico City - Guadalajara
- *El Rápido de la Frontera* (railcar service) Chihuahua - Ciudad Juárez
- *El Hidalguense* - Mexico City - Pachuca, Hidalgo.

Other passenger service was provided between Mexico City and: Cuernavaca, Morelos; Tampico, Tamaulipas; Guanajuato, Guanajuato; and Veracruz, Veracruz.

See also

- List of Mexican railroads

References

[1] Minsk, R. Todd (July 2003). "Part 1: 1870-1907" (http://www.dartmouth.edu/~rtminsk/queretarotracks1.html). *Querétaro, Qro.: a chronology of railroad development.* . Retrieved March 7, 2009.
[2] Ferrocariles Nacionales de México (September 10, 2008). "Ferrocariles Nacionales de México en Liquidación" (http://www.fnm.com.mx). . Retrieved March 7, 2009.

Privatization

Privatization is the incidence or process of transferring ownership of a business, enterprise, agency or public service from the public sector (the state or government) to the private sector (businesses that operate for a private profit) or to private non-profit organizations. The term is also used in a quite different sense, to mean government out-sourcing of services to private firms, e.g. functions like revenue collection, law enforcement, and prison management.[1]

The term "privatization" also has been used to describe two unrelated transactions. The first is a buyout, by the majority owner, of all shares of a public corporation or holding company's stock, privatizing a publicly traded stock, and often described as private equity. The second is a demutualization of a mutual organization or cooperative to form a joint stock company.[2]

Privatisation generally improves the output and efficiency of the organisations that are privatised.[3]

Origin

Edwards states that *The Economist* coined the term in the 1930s in covering Nazi German economic policy.[4] [5]

The Oxford English Dictionary notes usage dating from 1942 in *Econ. Jrnl*, 52, 398.

History

A long history of privatization dates from Ancient Greece, when governments contracted out almost everything to the private sector.[6] In the Roman Republic private individuals and companies performed the majority of services including tax collection (tax farming), army supplies (military contractors), religious sacrifices and construction. However, the Roman Empire also created state-owned enterprises—for example, much of the grain was eventually produced on estates owned by the Emperor. Some scholars suggest that the cost of bureaucracy was one of the reasons for the fall of the Roman Empire.[6]

Perhaps one of the first ideological movements towards privatization came during China's golden age of the Han dynasty. Taoism came into prominence for the first time at a state level, and it advocated the laissez-faire principle of Wu wei (), literally meaning "do nothing".[7] The rulers were counseled by the Taoist clergy that a strong ruler was virtually invisible.

During the Renaissance, most of Europe was still by and large following the feudal economic model. By contrast, the Ming dynasty in China began once more to practice privatization, especially with regards to their manufacturing industries. This was a reversal of the earlier Song dynasty policies, which had themselves overturned earlier policies in favor of more rigorous state control.[8]

In Britain, the privatization of common lands is referred to as enclosure (in Scotland as the Lowland Clearances and the Highland Clearances). Significant privatizations of this nature occurred from 1760 to 1820, coincident with the industrial revolution in that country.

In more recent times, Winston Churchill's government privatized the British steel industry in the 1950s, and West Germany's government embarked on large-scale privatization, including selling its majority stake in Volkswagen to small investors in a public share offering in 1961.[6] In the 1970s General Pinochet implemented a significant privatization program in Chile. However, it was in the 1980s under the leaderships of Margaret Thatcher in the UK and Ronald Reagan in the USA, that privatization gained worldwide momentum. In the UK this culminated in the 1993 privatization of British Rail under Thatcher's successor, John Major; British Rail having been formed by prior nationalization of private rail companies.

Significant privatization of state owned enterprises in Eastern and Central Europe and the former Soviet Union was undertaken in the 1990s with assistance from the World Bank, the U.S. Agency for International Development, the German Treuhand, and other governmental and nongovernmental organizations.

A major ongoing privatization, that of Japan Post, involves the Japanese post service and the largest bank in the world. This privatization, spearheaded by Junichiro Koizumi, started in 2007 following generations of debate. The privatization process is expected to last until 2017.

Types

There are four main methods of privatization:

1. Share issue privatization (SIP) - selling shares on the stock market
2. Asset sale privatization - selling an entire organization (or part of it) to a strategic investor, usually by auction or by using the Treuhand model
3. Voucher privatization - distributing shares of ownership to all citizens, usually for free or at a very low price.
4. Privatization from below - Start-up of new private businesses in formerly socialist countries.

Choice of sale method is influenced by the capital market, political and firm-specific factors. SIPs are more likely to be used when capital markets are less developed and there is lower income inequality. Share issues can broaden and deepen domestic capital markets, boosting liquidity and (potentially) economic growth, but if the capital markets are insufficiently developed it may be difficult to find enough buyers, and transaction costs (e.g. underpricing required) may be higher. For this reason, many governments elect for listings in the more developed and liquid markets, for example Euronext, and the London, New York and Hong Kong stock exchanges.

As a result of higher political and currency risk deterring foreign investors, asset sales occur more commonly in developing countries.

Voucher privatization has mainly occurred in the transition economies of Central and Eastern Europe, such as Russia, Poland, the Czech Republic, and Slovakia. Additionally, Privatization from below is/has been an important type of economic growth in transition economies.

A substantial benefit of share or asset-sale privatizations is that bidders compete to offer the highest price, creating income for the state in addition to tax revenues. Voucher privatizations, on the other hand, could be a genuine transfer of assets to the general population, creating a real sense of participation and inclusion. If the transfer of vouchers is permitted, a market in vouchers could be created, with companies offering to pay money for them.

Results

Literature reviews[9] [10] find that in competitive industries with well-informed consumers, privatization consistently improves efficiency. The more competitive the industry, the greater the improvement in output, profitability, and efficiency.[3] Such efficiency gains mean a one-off increase in GDP, but through improved incentives to innovate and reduce costs also tend to raise the rate of economic growth. The type of industries to which this generally applies include manufacturing and retailing. Although typically there are social costs associated with these efficiency gains, [11] many economists argue that these can be dealt with by appropriate government support through redistribution and perhaps retraining.

In sectors that are natural monopolies or public services (such as, say, passenger rail in the United States), the results of privatization are much more mixed, as a private monopoly behaves much the same as a public one in liberal economic theory. The government is actually seen as a more natural provider of public goods and services. However, the efficiency of an existing public sector operation can be put into question requiring changes to be made. Changes may include, inter alia, the imposition of related reforms such as greater transparency and accountability of management, an improved cost-benefit analysis, improved internal controls, regulatory systems, and better financing, rather than privatization itself.

Regarding political corruption, it is a controversial issue whether the size of the public sector per se results in corruption. The Nordic countries have low corruption but large public sectors. However, these countries score high on the Ease of Doing Business Index, due to good and often simple regulations, and for political rights and civil liberties, showing high government accountability and transparency. One should also notice the successful, corruption-free privatizations and restructuring of government enterprises in the Nordic countries. For example, dismantling telecommunications monopolies has resulted in several new players entering the market and intense competition with price and service.

Also regarding corruption, the sales themselves give a large opportunity for grand corruption. Privatizations in Russia and Latin America were accompanied by large-scale corruption during the sale of the state-owned companies. Those with political connections unfairly gained large wealth, which has discredited privatization in these regions. While media have reported widely the grand corruption that accompanied the sales, studies have argued that in addition to increased operating efficiency, daily petty corruption is, or would be, larger without privatization, and that corruption is more prevalent in non-privatized sectors. Furthermore, there is evidence to suggest that extralegal and unofficial activities are more prevalent in countries that privatized less.[12]

Differing views

Supporting

Studies show that private market factors can more efficiently deliver many goods or service than governments due to free market competition.[3] [9] [10] Over time this tends to lead to lower prices, improved quality, more choices, less corruption, less red tape, and/or quicker delivery. Many proponents do not argue that everything should be privatized. According to them, market failures and natural monopolies could be problematic. However, anarcho-capitalists prefer that every function of the state be privatized, including defense and dispute resolution.

The basic economic argument given for privatization states that governments have few incentives to ensure that the enterprises they own are well run. One problem is the lack of comparison in state monopolies. It is difficult to know if an enterprise is efficient or not without competitors to compare against. Another is that the central government administration, and the voters who elect them, have difficulty evaluating the efficiency of numerous and very different enterprises. A private owner, often specializing and gaining great knowledge about a certain industrial sector, can evaluate and then reward or punish the management in much fewer enterprises much more efficiently. Also, governments can raise money by taxation or simply printing money should revenues be insufficient, unlike a private owner.

If private and state-owned enterprises compete against each other, then the state owned may borrow money more cheaply from the debt markets than private enterprises, since the state owned enterprises are ultimately backed by the taxation and printing press power of the state, gaining an unfair advantage.

Privatizing a non-profitable state-owned company may force the company to raise prices in order to become profitable. However, this would remove the need for the state to provide tax money in order to cover the losses.

Proponents of privatization make the following arguments:

- Performance. State-run industries tend to be bureaucratic. A political government may only be motivated to improve a function when its poor performance becomes politically sensitive, and such an improvement can be reversed easily by another regime.
- Increased efficiency. Private companies and firms have a greater incentive to produce more goods and services for the sake of reaching a customer base and hence increasing profits. A public organization would not be as productive due to the lack of financing allocated by the entire government's budget that must consider other areas of the economy.
- Specialization. A private business has the ability to focus all relevant human and financial resources onto specific functions. A state-owned firm does not have the necessary resources to specialize its goods and services as a result of the general products provided to the greatest number of people in the population.
- Improvements. Conversely, the government may put off improvements due to political sensitivity and special interests—even in cases of companies that are run well and better serve their customers' needs.

- Corruption. A state-monopolized function is prone to corruption; decisions are made primarily for political reasons, personal gain of the decision-maker (i.e. "graft"), rather than economic ones. Corruption (or principal-agent issues) in a state-run corporation affects the ongoing asset stream and company performance, whereas any corruption that may occur during the privatization process is a one-time event and does not affect ongoing cash flow or performance of the company.
- Accountability. Managers of privately owned companies are accountable to their owners/shareholders and to the consumer, and can only exist and thrive where needs are met. Managers of publicly owned companies are required to be more accountable to the broader community and to political "stakeholders". This can reduce their ability to directly and specifically serve the needs of their customers, and can bias investment decisions away from otherwise profitable areas.
- Civil-liberty concerns. A company controlled by the state may have access to information or assets which may be used against dissidents or any individuals who disagree with their policies.
- Goals. A political government tends to run an industry or company for political goals rather than economic ones.
- Capital. Privately held companies can sometimes more easily raise investment capital in the financial markets when such local markets exist and are suitably liquid. While interest rates for private companies are often higher than for government debt, this can serve as a useful constraint to promote efficient investments by private companies, instead of cross-subsidizing them with the overall credit-risk of the country. Investment decisions are then governed by market interest rates. State-owned industries have to compete with demands from other government departments and special interests. In either case, for smaller markets, political risk may add substantially to the cost of capital.
- Security. Governments have had the tendency to "bail out" poorly run businesses, often due to the sensitivity of job losses, when economically, it may be better to let the business fold.
- Lack of market discipline. Poorly managed state companies are insulated from the same discipline as private companies, which could go bankrupt, have their management removed, or be taken over by competitors. Private companies are also able to take greater risks and then seek bankruptcy protection against creditors if those risks turn sour.
- Natural monopolies. The existence of natural monopolies does not mean that these sectors must be state owned. Governments can enact or are armed with anti-trust legislation and bodies to deal with anti-competitive behavior of all companies public or private.
- Concentration of wealth. Ownership of and profits from successful enterprises tend to be dispersed and diversified -particularly in voucher privatization. The availability of more investment vehicles stimulates capital markets and promotes liquidity and job creation.
- Political influence. Nationalized industries are prone to interference from politicians for political or populist reasons. Examples include making an industry buy supplies from local producers (when that may be more expensive than buying from abroad), forcing an industry to freeze its prices/fares to satisfy the electorate or control inflation, increasing its staffing to reduce unemployment, or moving its operations to marginal constituencies.
- Profits. Corporations exist to generate profits for their shareholders. Private companies make a profit by enticing consumers to buy their products in preference to their competitors' (or by increasing primary demand for their products, or by reducing costs). Private corporations typically profit more if they serve the needs of their clients well. Corporations of different sizes may target different market niches in order to focus on marginal groups and satisfy their demand. A company with good corporate governance will therefore be incentivized to meet the needs of its customers efficiently.
- Job gains. As the economy becomes more efficient, more profits are obtained and no government subsidies and less taxes are needed, there will be more private money available for investments and consumption and more profitable and better-paid jobs will be created than in the case of a more regulated economy.[13]

Opposing

Opponents of certain privatizations believe that certain public goods and services should remain primarily in the hands of government in order to ensure that everyone in society has access to them (such as law enforcement, basic health care, and basic education). Likewise, private goods and services should remain in the hands of the private sector. There is a positive externality when the government provides public goods and services to society at large, such as defense and disease control. As for natural monopolies, opponents of privatisation claim that they are subject to fair competition, and better administrated by the state.

Many privatization opponents also warn against the practice's inherent tendency toward corruption. As many areas which the government could provide are essentially profitless, the only way private companies could, to any degree, operate them would be through contracts or block payments. In these cases, the private firm's performance in a particular project would be removed from their performance, and embezzlement and dangerous cost-cutting measures might be taken to maximize profits.

Some would also point out that privatizing certain functions of government might hamper coordination, and charge firms with specialized and limited capabilities to perform functions which they are not suited for. In rebuilding a war torn nation's infrastructure, for example, a private firm would, in order to provide security, either have to hire security, which would be both necessarily limited and complicate their functions, or coordinate with government, which, due to a lack of command structure shared between firm and government, might be difficult. A government agency, on the other hand, would have the entire military of a nation to draw upon for security, whose chain of command is clearly defined. Opponents would say that this is a false assertion: numerous books refer to poor organization between government departments (for example the Hurricane Katrina incident).

Although private companies will provide a similar good or service alongside the government, opponents of privatization are careful about completely transferring the provision of public goods, services and assets into private hands for the following reasons:

- Performance. A democratically elected government is accountable to the people through a legislature, Congress or Parliament, and is motivated to safeguarding the assets of the nation. The profit motive may be subordinated to social objectives.
- Improvements. the government is motivated to performance improvements as well run businesses contribute to the State's revenues.
- Corruption. Government ministers and civil servants are bound to uphold the highest ethical standards, and standards of probity are guaranteed through codes of conduct and declarations of interest. However, the selling process could lack transparency, allowing the purchaser and civil servants controlling the sale to gain personally.
- Accountability. The public does not have any control or oversight of private companies.
- Civil-liberty concerns. A democratically elected government is accountable to the people through a parliament, and can intervene when civil liberties are threatened.
- Goals. The government may seek to use state companies as instruments to further social goals for the benefit of the nation as a whole.
- Capital. Governments can raise money in the financial markets most cheaply to re-lend to state-owned enterprises.
- Strategic and Sensitive areas. Governments have chosen to keep certain companies/industries under public control because of their strategic importance or sensitive nature.
- Cuts in essential services. If a government-owned company providing an essential service (such as the water supply) to all citizens is privatized, its new owner(s) could lead to the abandoning of the social obligation to those who are less able to pay, or to regions where this service is unprofitable.
- Natural monopolies. Privatization will not result in true competition if a natural monopoly exists.
- Concentration of wealth. Profits from successful enterprises end up in private, often foreign, hands instead of being available for the common good.

- Political influence. Governments may more easily exert pressure on state-owned firms to help implementing government policy.
- Downsizing. Private companies often face a conflict between profitability and service levels, and could over-react to short-term events. A state-owned company might have a longer-term view, and thus be less likely to cut back on maintenance or staff costs, training etc., to stem short term losses. Many private companies have downsized while making record profits.
- Profit. Private companies do not have any goal other than to maximize profits. A private company will serve the needs of those who are most willing (and able) to pay, as opposed to the needs of the majority, and are thus anti-democratic. The more necessary a good is, the lower the price elasticity of demand, as people will attempt to buy it no matter the price. In the case of price elasticity of demand is zero (perfectly inelastic good), demand part of supply and demand theories does not work.
- Privatization and Poverty. It is acknowledged by many studies that there are winners and losers with privatization. The number of losers —which may add up to the size and severity of poverty—can be unexpectedly large if the method and process of privatization and how it is implemented are seriously flawed (e.g. lack of transparency leading to state-owned assets being appropriated at minuscule amounts by those with political connections, absence of regulatory institutions leading to transfer of monopoly rents from public to private sector, improper design and inadequate control of the privatization process leading to asset stripping.[14]
- Job Loss. Due to the additional financial burden placed on privatized companies to succeed without any government help, unlike the public companies, jobs could be lost to keep more money in the company.

Equivalence to secured borrowing

Setting aside questions of efficiency and public versus private control of resources, some privatization transactions can be interpreted as a form of a secured loan,[15] [16] and are criticized as a "particularly noxious form of governmental debt".[15] In this interpretation, the upfront payment from the privatization sale corresponds to the principal amount of the loan, while the proceeds from the underlying asset correspond to secured interest payments – the transaction can be considered substantively the same as a secured loan, though it is structured as a sale.[15] This interpretation is particularly argued to apply to recent municipal transactions in the United States, particularly for fixed term, such as the 2008 sale of the proceeds from Chicago parking meters for 75 years. It is argued that this is motivated by "politicians' desires to borrow money surreptitiously",[15] due to legal restrictions on and political resistance to alternative sources of revenue, viz, raising taxes or issuing debt.

Intermediate views

Others don't dispute that well-run for-profit entities with sound corporate governance may be considerably more efficient than an inefficient governmental bureaucracy or NGO, however many **implementations** of privatization can - in practice - lead to the fire sale of public assets, and/or to inefficient or corrupt - for profit management.

Developed or minimally corrupt economies

A top executive can readily **reduce** the perceived value of an asset – due to information asymmetry. The executive can accelerate accounting of expected expenses, delay accounting of expected revenue, engage in off balance sheet transactions to make the company's profitability appear temporarily poorer, or simply promote and report severely conservative (e.g. pessimistic) estimates of future earnings. Such seemingly adverse earnings news will be likely to (at least temporarily) reduce sale price. (This is again due to information asymmetries since it is more common for top executives to do everything they can to window dress their earnings forecasts). There are typically very few legal risks to being 'too conservative' in one's accounting and earnings estimates.

When the entity gets taken private - at a dramatically lower price - the new private owner gains a windfall from the former top executive's actions to (surreptitiously) reduce the sales price. This can represent tens of billions of dollars (questionably) transferred from previous owners (the public) to the takeover artist. The former top executive is then rewarded with a golden handshake for presiding over the fire sale that can sometimes be in the tens or hundreds of millions of dollars for one or two years of work. (This is nevertheless an excellent bargain for the takeover artist, who will tend to benefit from developing a reputation of being very generous to parting top executives).

When a publicly held asset, mutual or non-profit organization undergoes privatization, top executives often reap tremendous monetary benefits. The executives can facilitate the process by making the entity appear to be in financial crisis - this reduces the sale price (to the profit of the purchaser), and makes non-profits and governments more likely to sell.

Ironically, it can also contribute to a public perception that private entities are more efficiently run reinforcing the political will to sell of public assets. Again, due to asymmetric information, policy makers and the general public see a government owned firm that was a financial 'disaster' - miraculously turned around by the private sector (and typically resold) within a few years.

Underdeveloped or highly corrupt economies

In a society with substantial corruption, privatization allows the government currently in power and its backers to siphon a large portion of the entire net present value of state assets away from the public and into the accounts of their favored power brokers. Without privatization, corrupt officials would have to slowly harvest their corrupt earnings over time. As such, efficient privatization depends on their being a very low of **current** corruption among the current government officials since it allows for far more 'efficient' extraction of corrupt rents.

Of course, corrupt governments can also extract corrupt rents quite efficiently in other ways - particularly by borrowing extensively to engage in spending on overly favorable contracts with their backers (or on tax shelters, subsidies or other giveaways). Generations of subsequent taxpayers are then left with paying back the debt incurred for corrupt transfers made decades previously. Naturally, this may lead to the sale of public assets....

In the end, the public is left with a government that taxes them heavily, and gives them nothing in return. Debt repayment is enforced by international agreements and agencies such as the IMF. Infrastructure and upkeep is sacrificed - leading to a further decay in the economic efficiency of the country over time.

Alternatives

Public utility

The enterprise can remain as a public utility.

Non-profit

A private non-profit organization could manage the enterprise.

Municipalization

Transferring control to municipal government

Outsourcing or sub-contracting

National services may sub-contract or out-source functions to private enterprises. A notable example of this is in the United Kingdom, where many municipalities have contracted out their garbage collection or administration of parking fines to private companies. In addition, the British government has involved the private sector more in the workings of the National Health Service principally through outsourcing the construction and operation of new hospitals to private companies. There are also moves to refer patients to private surgeries to ease the load on existing NHS human resources, and covering the cost of this.

Partial ownership

An enterprise may be privatized, but with the state retaining a number of shares in the resultant company. This is a particularly notable phenomenon in France, where the state often retains a "blocking stake" in private industries. In Germany, the state privatized Deutsche Telekom in small tranches, and still retains about a third of the company. As of 2005, the state of North Rhine-Westphalia is also planning to buy shares in the energy company E.ON in what is claimed to be an attempt to control spiraling costs.

Whilst partial privatization could be an alternative, it is more often a stepping stone to full privatization. It can offer the business a smoother transition period during which it can gradually adjust to market competition. Some state-owned companies are so large that there is the risk of sucking liquidity from the rest of the market, even in the most liquid marketplaces: this may favor gradual privatization. The first *tranche* of a multi-step privatization would also in the first instance establish a valuation for the enterprise to mitigate complaints of under-pricing.

In some instances of partial privatization of contracted services, some portion(s) of the state-owned service are provided by private-sector contractors, but the government retains the capacity to self-operate at contract intervals, if it so chooses. An example of partial privatization would be some forms of school bus service contracting, such as arrangements where equipment and other resources purchased with government capital funds and/o those already owned by a governmental entity are used by the contractor for a period of time in providing services, but ownership is retained by the governmental unit. This form of partial privatization eases concerns that once an operation is contracted, the government may be unable to obtain sufficient competitive bids, and be subjected to terms less desirable than the prior operation under state-ownership. Under that scenario, a *reverse privatization* would be more feasible for the government. (see section below)

Notable examples

The largest privatization in history involved Japan Post. It was the nation's largest employer and one third of all Japanese government employees worked for Japan Post. Japan Post was often said to be the largest holder of personal savings in the world.

The Prime Minister Junichiro Koizumi wanted to privatize it because it was thought to be an inefficient and a source for corruption. In September 2003, Koizumi's cabinet proposed splitting Japan Post into four separate companies: a bank, an insurance company, a postal service company, and a fourth company to handle the post offices as retail storefronts of the other three.

After the Upper House rejected privatization, Koizumi scheduled nationwide elections for September 11, 2005. He declared the election to be a referendum on postal privatization. Koizumi subsequently won this election, gaining the necessary supermajority and a mandate for reform, and in October 2005, the bill was passed to privatize Japan Post in 2007.[17]

Nippon Telegraph and Telephone's privatization in 1987 involved the largest share-offering in financial history at the time.[18] 15 of the world's 20 largest public share offerings have been privatizations of telecoms.[18]

The United Kingdom's largest public-share offerings were privatizations of British Telecom and British Gas during the 1980s under the Conservative government of Margaret Thatcher, when many state-run firms were sold off to the private sector. This attracted very mixed views from the public and parliament, and even a former Conservative prime minister, Harold Macmillan, was critical of the policy; likening it to "selling the family silver".[19]

The largest public-share offering in France was France Telecom.

Egypt undertook widespread privatization under President Hosni Mubarak. After his overthrow in the 2011 revolution, the association of the newly private businesses with the crony capitalism of the old regime along with the new look at long-festering labor and police-state issues have led to calls for re-nationalization.[20]

Negative responses

Privatization proposals in key public service sectors such as water and electricity in many cases meet with strong resistance from opposition political parties and from civil society groups, many of which regard them as natural monopolies. Campaigns typically involve demonstrations and democratic political activities; sometimes the authorities attempt to suppress opposition using violence (e.g. Cochabamba protests of 2000 in Bolivia and protests in Arequipa, Peru, in June 2002). Opposition is often strongly supported by trade unions. Opposition is usually strongest to water privatization—as well as Cochabamba, recent examples include Haiti, Ghana and Uruguay (2004). In the latter case a civil-society-initiated referendum banning water privatization was passed in October 2004.

Reversion

A reversion from contracted ownership of an enterprise or services to governmental ownership and/or provision is called *reverse privatization* or nationalization. Such a situation most often occurs when a privatization contractor fails financially and/or the governmental unit has failed to purchase satisfactory service at prices it regards as less than with state-ownership or self-operation of services. Another circumstance may occur when greater control than viable under privatization is determined to be in the governmental unit's best interest.

National-security concerns may be the source of reverse privatization actions when the most likely providers are non-domestic or international corporations or entities. For example, in 2001, in response to the September 11th attacks, the then-private airport security industry in the United States was nationalized and put under the authority of the Transportation Security Administration.

See also

- Corporatization
- Deregulation
- Marketization
- Public ownership
- Securitization
- Too Big to Fail
- Welfare state

Case studies:

- Privatization in Russia
- Privatization of British Rail
- Privatization of public toilets

Development strategies:

- Private sector development
- Special Economic Zone
- Urban Enterprise Zone

Notes

[1] Chowdhury, F. L. "Corrupt Bureaucracy and Privatisation of Tax Enforcement", 2006: Pathak Samabesh, Dhaka.

[2] "Musselburgh Co-op in crisis as privatization bid fails." (http://www.thenews.coop/index.php?content=story&id=835). Co-operative News. 2005-11-01. . Retrieved 2008-05-21.

[3] "Privatising State-owned Enterprises" (http://www.apec.org.au/docs/10_TP_PFI 4/Privatising SOEs.pdf). 2010-02-22. p. 9. . Retrieved 2011-07-11.

[4] Edwards, Ruth Dudley (1995). *The Pursuit of Reason: The Economist 1843-1993*. Harvard Business School Press. p. 946. ISBN 0-87584-608-4.

[5] Compare Bel, Germà (2006). "Retrospectives: The Coining of 'Privatisation' and Germany's National Socialist Party". *Journal of Economic Perspectives* **20** (3): 187–194. doi:10.1257/jep.20.3.187.

[6] *International Handbook on Privatization* by David Parker, David S. Saal

[7] Li & Zheng 2001, p. 241

[8] Bouye, Thomas M., Manslaughter, markets, and moral economy

[9] "Privatisation in Competitive Sectors: The Record to Date, World Bank Policy Research Working Paper No. 2860". *John Nellis and Sunita Kikeri* (World Bank). June 2002. SSRN 636224.

[10] "From State To Market: A Survey Of Empirical Studies On Privatisation" (http://faculty-staff.ou.edu/M/William.L.Megginson-1/prvsvpapJLE.pdf) (PDF). *William L. Megginson and Jeffry M. Netter* (*Journal of Economic Literature*). *June 2001.* .

[11] "Winners and Losers: Assessing the Distributional Impact of Privatisation, CGD Working Paper No 6" (http://www.cgdev.org/docs/cgd_wp006.pdf) (PDF). *Nancy Birdsall & John Nellis* (Center for Global Development). March 9, 2006. .

[12] Privatisation in Competitive Sectors: The Record to Date. Sunita Kikeri and John Nellis. World Bank Policy Research Working Paper 2860, June 2002. Privatisation and Corruption (http://idei.fr/doc/conf/veol/straub_martimort.pdf). David Martimort and Stéphane Straub.

One career city manager in America, Roger L. Kemp, wrote a library reference volume titled *Privatization: The Provision of Public Services by the Private Sector," which was originally published in 1991 and republished in 2007. In this volume, based on a national literature search of best practices among municipal governments in this field, Dr. Kemp recommended that administrators owe it to their taxpayers and citizens to seek private alternatives to selected public services. He felt that city managers should go to the marketplace to determine the cost of contracting for selected public services, while keeping quality consistent with the same service provided by the municipality. Sometimes, Kemp noted, it's more cost effective to have certain public services provided by contract by the private sector. In some cases it can be less expensive for the private sector to provide public services, but the benefit to society may actually turn out to be less as well.*

[13] Central Europe's Mass-Production Privatization (http://www.heritage.org/research/europe/HL352.cfm), Heritage Lecture #352

[14] Dagdeviren (2006) "Revisiting privatisation in the context of poverty alleviation" Journal of International Development, Vol. 18, 469–488

[15] Roin, Julie. "Privatization and the Sale of Tax Revenues" (http://papers.ssrn.com/sol3/papers.cfm?abstract_id=1880033). *SSRN eLibrary.* . Retrieved 2011-07-27, also published as " Privatization and the Sale of Tax Revenues (http://www.minnesotalawreview.org/articles/privatization-sale-tax-revenues/)" in *Minnesota Law Review* (http://www.minnesotalawreview.org/), Vol. 85, p. 1965, 2011, and *U of Chicago Law & Economics,* Olin Working Paper (http://www.law.uchicago.edu/lawecon/workingpapers) No. 560

[16] U. of C. professor argues privatization of public assets just like borrowing money (http://articles.chicagotribune.com/2011-07-22/business/ct-biz-0722-chicago-law-20110722_1_parking-meters-privatization-summer-reading), July 22, 2011, *Chicago Tribune,* Ameet Sachdev's Chicago Law, Ameet Sachdev

[17] Takahara, "All eyes on Japan Post"Faiola, Anthony (2005-10-15). "Japan Approves Postal Privatization" (http://www.washingtonpost.com/wp-dyn/content/article/2005/10/14/AR2005101402163.html). *Washington Post* (The Washington Post Company): p. A10. . Retrieved 2007-02-09.

[18] The Financial Economics of Privatisation By William L. Megginson, p. 205 - 206

[19] (http://www.number10.gov.uk/history-and-tour/harold-macmillan-2/)

[20] Amos, Deborah, "In Egypt, Revolution Moves Into The Factories" (http://www.npr.org/2011/04/20/135542498/in-egypt-revolution-moves-into-the-factories), *NPR*, April 20, 2011. Retrieved 2011-04-20.

References

* Alexander, Jason. 2009. Contracting Through the Lens of Classical Pragmatism: An Exploration of Local Government Contracting. Applied Research Project. Texas State University. http://ecommons.txstate.edu/arp/288/.
* Dovalina, Jessica. 2006. Assessing the Ethical Issues Found in the Contracting Out Process. Applied Research Project. Texas State University. http://ecommons.txstate.edu/arp/108/.
* Segerfeldt, Fredrik. 2006. Water for sale: how business and the market can resolve the world's water crisis. Stockholm Network. http://www.stockholm-network.org/downloads/events/d41d8cd9-Amigo%20Segerfeldt.pdf
* Bernard Black et al., 'Russian Privatization and Corporate Governance: What Went Wrong? (2000) 52 Stanford Law Review 1731

Unindexed

* Zullo, Roland. (2009). Does Fiscal Stress Induce Privatization? Correlates of Private and Intermunicipal Contracting, 1992-2002 (http://www3.interscience.wiley.com/journal/122466875/abstract). Governance 22.3 (July): 459-481.
* Kosar, Kevin R. (2006), "Privatisation and the Federal Government: An Introduction" (http://www.fas.org/sgp/crs/misc/RL33777.pdf), *Report from the Congressional Research Service*
* Bel, Germà (2006), "The coining of `privatisation´and Germany's National Socialist Party" (http://www.ub.es/graap/JEP.pdf), *Journal of Economic Perspectives* 20(3), 187-194
* Clarke, Thomas (ed.) (1994) "International Privatisation: Strategies and Practices" Berlin and New York: Walter de Gruyter, ISBN 3-11-013569-8
* Clarke, Thomas and Pitelis, Christos (eds.) (1995) "The Political Economy of Privatisation" London and New York: Routledge, ISBN 0-415-12705-X
* Nico Perrone (2002), *Economia pubblica rimossa*, Milan, Giuffrè ISBN 88-14-10088-8
* Mayer, Florian (2006) *Vom Niedergang des unternehmerisch tätigen Staates: Privatisierungspolitik in Großbritannien, Frankreich, Italien und Deutschland*, VS Verlag, Wiesbaden, ISBN 3-531-14918-0
* Megginson and Netter, From state to market: A survey of empirical studies on privatization,

Journal of Economic Literature 39(2), June 2001, 321-89.

* Juliet D'Souza, William L. Megginson (1999), "The Financial and Operating Performance of Privatised Firms during the 1990s" (http://faculty-staff.ou.edu/M/William.L.Megginson-1/prv90pap.pdf), *Journal of Finance* August 1999
* von Hayek, Friedrich, (1960) *The Constitution of Liberty*
* Smith, Adam (1994) *The Wealth of Nations*
* Stiglitz, Joseph *Globalization and its Discontents*
* David T. Beito, Peter Gordon, and Alexander Tabarrok (editors); foreword by Paul Johnson (2002). *The voluntary city: choice, community, and civil society*. Ann Arbor: University of Michigan Press/The Independent Institute. ISBN 0-472-08837-8.
* von Weizsäcker, Ernst, Oran Young, and Matthias Finger (editors): Limits to Privatisation. Earthscan, London 2005 ISBN 1-84407-177-4
* Jeb Sprague, 2007. Haiti: Workers Protest Privatisation Layoffs (http://ipsnews.net/news.asp?idnews=38646). Inter Press Service.
* Roger L. Kemp, PhD, "Privatization: The Provision of Public Services by the Private Sector," McFarland & Co., Jefferson, NC, USA; and London, UK., 2007.

External links

- In the Public Interest (http://www.inthepublicinterest.org) is a Resource Center on privatization and responsible contracting.
- Reports of the Public Services International Research Unit at the University of Greenwich (http://www.psiru. org/publicationsindex.asp) Research database with many articles on the effects of privatization
- Parker, David. "Privatisation ten years on : a critical analysis of its rationale and results" (http://dspace.lib. cranfield.ac.uk/handle/1826/606). Cranfield University, School of Management.

Ferrosur

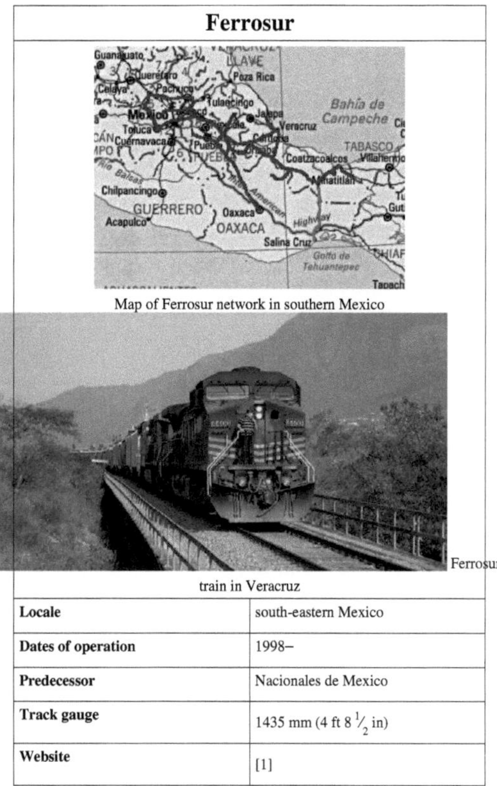

Ferrosur
train in Veracruz

Locale	south-eastern Mexico
Dates of operation	1998–
Predecessor	Nacionales de Mexico
Track gauge	1435 mm (4 ft 8 $\frac{1}{2}$ in)
Website	[1]

The **Ferrocarril del Sureste** (reporting mark **FSRR**), commonly known as Ferrosur, is a railroad that serves the south-eastern regions of Mexico.

The company was formed by the granting of two concessions as part of the privatization of Mexico's railways in 1998 and the subsequent merger of those two concessions in 2000. The company operates the vital rail line between Mexico City and Mexico's busiest Gulf of Mexico/Atlantic port at Veracruz.

Merger with Ferromex

Ferrosur was owned by Grupo Carso and Grupo Financiero Inbursa (companies controlled by Carlos Slim Helu) until November 2005 when the company was sold to Grupo Mexico, owner of Ferromex, in a US$309 million stock transaction.[2] The Mexican Federal Competition Commission (CFC) had rejected a proposed 2002 merger of Ferromex and Ferrosur amid opposition from Ferromex competitor, Grupo Transportación Ferroviaria Mexicana (TFM).[3]

Following the November 2005 purchase of Ferrosur by Grupo Mexico, Kansas City Southern de México (KCSM), successor to TFM, petitioned the Mexican government to block the merger of Ferrosur and Ferromex. The CFC rejected the merger in June 2006 stating that the merger would have led to excessive concentration in the railroad industry to the detriment of consumers and competing shippers.[4] However, in March 2011, a tribunal ruled in Grupo Mexico's favor, and the merger was permitted.[5]

See also

- List of Mexican railroads
- Rail transport in Mexico

Gallery

"La Soledad" Bridge (February 23, 1909)

Notes

[1] http://www.ferrosur.com.mx/
[2] FWN Select, "Grupo Mexico Buys Ferrosur Railway From Carso" (Nov. 25, 2005)
[3] FWN Select, "Mexico's Antitrust Agency To Study Railway Merger" (November 28, 2005)
[4] FWN Select, "Mexico's Antitrust Commission Rejects Rail Merger Appeal" (Nov. 15, 2006)
[5] "Mexican Tribunal OKs Grupo Mexico Railroad merger" (http://af.reuters.com/article/metalsNews/idAFN2821457620110328). Reuters. 2011-3-28. .

External links

- FERROSUR Official Site (http://www.ferrosur.com.mx/gxpsites/hgxpp001.aspx)
- MEXLIST—The Group for Mexican Railway Information (http://mexlist.com)

Article Sources and Contributors

Ferrocarril de Veracruz al Istmo *Source*: http://en.wikipedia.org/w/index.php?title=Ferrocarril_de_Veracruz_al_Istmo *Contributors*: AvicAWB, NE2

Nationalization *Source*: http://en.wikipedia.org/w/index.php?title=Nationalization *Contributors*: 16@r, AED, APHST, Achmelvic, Ahoerstemeier, Alansohn, Albacore60, Aleksandr Grigoryev, Altenmann, Andrejj, Andycjp, Atlantia, AxG, BWP1234, Barrylb, Barryob, BaseTurnComplete, Battlecry, Beano, BillMasen, Bobrayner, Cacafuego95, Calorus, Campbro, Cassowary, Chris the speller, ChronoSphere, Ckatz, Clawed, CliffC, Colchicum, Colonies Chris, CommonsDelinker, Cookiedog, Coolcaesar, Cpeter9, Czyrko, Dandelions, DanielRigal, Darth Sidious, David Robins, David.Monniaux, DavidArthur, Dcflyer, DeFacto, Deepred6502, Dhoyos, Dnfenner, DocendoDiscimus, Dodge1884, EAi, ELouka, Eeekster, Elwrucko, Emmisa, EqualRights, Exor674, Fconaway, Fursday, G-Man, Galoubet, Gazifikator, Gbjerga, GcSwRhIc, GeneralBelly, Geschichte, Glane23, GraemeLeggett, Grassynoel, Hairy Dude, Hellbus, HerrSchnapps, Hmains, IZAK, Ianare, IceDragon64, Indefatigable, Indiealtphreak, Interlingua, J.delanoy, JMK, JaGa, Jerryseinfeld, John Quiggin, Joseph Solis in Australia, Jrp55262, KAMiKAZOW, Kbthompson, Keith D, Keith111, Keyblade92, Kihjin, Kronak, La goutte de pluie, Lacrimosus, Lapsed Pacifist, LarsHolmberg, Lightmouse, Linuxlad, Little Professor, LokiiT, Lord of the Isles, Loremaster, Lost tourist, Luc Pepin, MBRZ48, MFlet1, Madhava 1947, Mahlon, Maltesedog, Marek69, MarkS, Martylunsford, Mauls, Maurreen, Mdkarazim, Melromero, Michaeldsuarez, Mike18xx, MilkTheImp, Mouramoor, Mschlindwein, Nadav1, Narjuko, Netwhizkid, Nick Cooper, Nightside eclipse, Nikodemos, Nk, Nomi887, Nono64, Notmyrealname, Nowa, Nurg, One last pharaoh, PB0305, Pcuser42, Pearle, Penrithguy, Petrb, Philip Stevens, Piotrus, Plasma east, Pol098, Portraitsketch, Protobaltoslav, R'n'B, RW Marloe, Ralhazzaa, RanEagle, Randroide, Rd232, Redrose64, Redvers, Retroneo, RexNL, ReyBrujo, Rlegaarden, Rlquall, Roche-Kerr, Rrburke, Rwendland, S.Örvarr.S, SPUI, SaliqKhan, Sam Hocevar, Samuel Webster, Sardanaphalus, Savetheoceans, Sbattles, Scepia, Schuhpuppe, Searchingfortao, Semmler, Shadowjams, Sigma 7, Simishag, Skabat169, Sky Harbor, Slady, SlimVirgin, Sole Soul, Squids and Chips, Stars4change, Stephenb, Stevertigo, SummerWithMorons, Super wiki editor, Sushiya, Tabletop, Textorus, Tgolden14, Timrollpickering, Tins128, Trygfe, TubularWorld, Twp, Typ932, Una Smith, Vaoverland, Vegetator, Vkyrt, Vushi, Washburnmav, Wlodzimierz, Woohookitty, Ww2censor, XXXpinoy777, Xt828, Yamamoto Ichiro, Yellowdesk, Yulia Romero, Zequist, Zleitzen, 317 anonymous edits

Rail transport *Source*: http://en.wikipedia.org/w/index.php?title=Rail_transport *Contributors*: *drew, -Majestic-, 1122334455, 12.234.49.xxx, 16@r, A bit iffy, Aaaallleex, Abby, Acela2038, Acprail, Acroterion, AdamW, Adashiel, Ae-a, Aesopos, Ahoerstemeier, Ahuskay, Aitias, Alansohn, Alco83, Aleator, AlexPlante, Alexdan loghin, Alextrevelian 006, Allreet, AllyUnion, Altenmann, Alucard (Dr.), Amakuru, AmyzzXX, Andreas Kaganov, Andrewlim1, Andrewpmk, Andy Dingley, Anlace, Anne McDermott, Antandrus, Anthony Appleyard, Armando, ArnoldReinhold, Arsenikk, Arthur Ellis, Aseemsjohri, Ashishbijwe24, Aude, Autonerd, Avenue X at Cicero, AvicAWB, AxelBoldt, Ayushrocks6, BD2412, BRG, Backifran, Bambuway, Basketball1776, Bastin, Bathrobe, Bebestbe, Beland, Berendale2, Besselfunctions, Beyond silence, Bhadani, Bhtpbank, Bigdumbdinosaur, Biggus wickus, Biker Biker, Bikramshergill, Bill37212, Birdhurst, Biscuittin, Bkell, Black Kite, Blehfu, Blimpguy, Bluerasberry, Bobo192, Bogdangiusca, Borgx, Brambo, Brhaspati, Brian Pearson, Brian evans, BruceDLimber, Bryan Derksen, Btornado, Bulleid Pacific, Burgher, Camboxer, Can't sleep, clown will eat me, CanadianLinuxUser, Canterberry, Casey56, Caseyjonz, Cazort, Cecropia, Celardore, Central Data Bank, Cessator, Chaleyer61, ChanceButler, Charlottesometimes, Cheekdog, Chevin, Chris55, Christian List, Chun-hian, Civil Engineer III, Claude girardin, Cluth, Cmapm, Cmichael, Colonies Chris, Complete fanatic, Conversion script, Coolhawks88, Coosbane, Corpx, Corti, Courcelles, CrossHouses, CryptoDerk, Cwolfsheep, Cynical, DAJF, DJ Sturm, Damirgraffiti, Dancter, Daniel C, Dark Shikari, DarkFireTaker, Das48, Daveswagon, David.Monniaux, DavidArthur, Ddstretch, Dealerofsalvation, Dearmstrong, Deb, Dellarb, Delldot, DennisJOBrien@yahoo.com, Dethme0w, Devin.phillips, DianeSelvy, DigbyDalton, Dnalrom123, Dnfenner, Doc glasgow, Download, Dp67, Dragomiloff, Dratman, Dsteckelberg, Duncharris, Dysprosia, ESkog, EarthFurst, EdJogg, Edgar181, El C, ElectraFlarefire, Emdx, Emperor Genius, Emsox, Engi08, Epbr123, Epolk, Equinox137, Eric-Wester, Exe, Expertz123, Ezhiukas 7, FT2, Faradayplank, Farjiadmi, Ferrierd, Feydey, Finavon, Finngall, Fixerofnatasha, FlavrSavr, Flux.books, Foodman, Fortdj33, Frankenpuppy, Fred Bauder, Fredericsalve, Frixa, Fudoreaper, Furrykef, Fyrael, G-Man, Gandydancer, Gary D, Gbtojun, Gdo01, Geoff NoNick, Gerfriedc, Geschichte, Ghaly, Globalsolidarity, Golbez, Grafen, Graham87, Grim23, Gsaup, Gtstricky, Gun Powder Ma, Guybrarian, Gwernol, Gökhan, Hairy Dude, Harjasusi, Haroonn1, Harpblaster, Harperbruce, Hclincha, HenryLi, Hephaestos, Hmrox, Hoo man, Hu, Hugh Manatee, Hulagutten, Hydriotaphia, Ianp1950, Ichol, Ishlan, J.delanoy, JCam, JNW, JRR Trollkien, JYolkowski, JackLumber, Jacob.jose, Jallason, Jameswilson, Japman6, Jarle fagerheim, Jax184, Jclemens, Jcuk, Jdcook, Jdforrester, Jeepday, Jeffakolb, Jeremy5561, Jevon, Jhonchris1981, Jim, Jimp, Jinty182, Jmorgan, JoeK, Joethespaz13, Joey Roe, John, JohnOwens, Johnasher, Jojit fb, Jonathan.s.kt, JorgeGG, Joshua, Jovianeye, Jpgordon, Jpk, Jrleighton, Jstreutker, Juggleandhope, JuliusAugustus, Jusjih, Juxo, KF, KVDP, Kahuzi, Kaibabsquirrel, Kalemika, Kalixiri, Katsp8, Kazvorpal, Kdar, Keeper76, Keidansky, Keithmall, Kelisi, Kelly Martin, Keotaman, Ketanpujare, Kingpin13, Kkmd, Krath, Krigjsman, Ksyrie, Kumioko, Kungfuadam, Kuru, L blue 1, La Pianista, Lafleure355, Lars T., LeaveSleaves, Leefletcher, LeoO3, Leonard G., Leptictidium, Levineps, Lightmouse, Lindahq, Lockesdonkey, Longhair, Lord Bob, Loren36, LorenzoB, Lost on belmont, Lost tourist, Lot29, Lovemehateme123, Luna Santin, MNboy, MPD01605, Madcoverboy, Madhero88, MakeChooChooGoNow, Mamawrites, Man vyi, Mangoe, Manop, Maranjc, MarsRover, Marshman, Matt von Furrie, Mattisse, MauriceJFox3, Maury Markowitz, Mav, Mayooranathan, Mdd4696, Megaman en m, Mervyn, Mgaved, Michael Hardy, MichaelBillington, Michaelduly, Mike Dill, Mikeasolomon, Mindspillage, Minesweeper, Mitul0520, MoO, Moebiusuibeom-en, Moncrief, Monedula, Morven, Moyogo, Murray Langton, N2e, NE2, Naddy, NameThatWorks, Nanana123456789, Nat Krause, Nerfer, NicholasJones, Nick Number, Nicolae Coman, Nk, Noexit, Noisy, Northumbrian, Nricardo, Nunh-huh, Oanabay04, Oceanhahn, Olahus, OlavN, Olivier, Omnedon, Ortolan88, PFHLai, PNRStreamliner, PReinie, Pak bahadur, Patoune08, Patrick, Paul Matthews, Paul W, PaulDrye, Perdelsky, Peter Horn, PeterEastern, Peterkingiron, Pgk, Phlebas, Piano non troppo, Pickle UK, Pietrow, Pilgaard, Pilotguy, Pinethicket, Pion, Piotrus, Playclever, Plutochaun, Pne, Pointer1, Pressforaction, ProhibitOnions, Prolog, Propaniac, Psublue23, Qertis, QuantumEngineer, Quintote, R'n'B, RFBailey, RTucker, Rafikk, Randall uob, Redrose64, Regregex, Rememberway, RevRagnarok, Reward, Rich Farmbrough, Rich4560, Rickyrab, Rigadoun, Rjensen, Rjmccall, Rjstott, Rjwilmsi, Rmhermen, RobHarding, Robert Merkel, Rods42, Rojomoke, Roke, Rrburke, Rror, Rsrikanth05, Runewiki777, S.K., SEWilco, SNIyer12, SPUI, ST47, Salamurai, Sam Hocevar, Sameer rana100, Samuel 69105, Samuell, Sandman30s, Sarah Waggoner, Sarah tonen, Satori Son, Sc147, ScAvenger lv, Scoot-Overload, Scoutersig, Seaphoto, SebbeSwe, Sentorment, Sergio.solar, Sgfaig, Sharadbob, SheepNotGoats, Sheldwich14, Signalhead, SilkTork, Silverjonny, SimonP, Skapur, Slambo, Smalljim, Smerlinare, Smurrayinchester, Snottily, Snowmanradio, Socipoet, Soliloquial, Sonett72, Sp33dyphil, Spagf5, Spangineer, Spearhead, Spellchecker, SqueakBox, Squids and Chips, SriMesh, Srtig, Starbois, Steam-loco, Stefan2, Steinberger, Stephen C. Carlson, Stragermont, Struway2, Stumpytrain, Sulzer55, Supt. of Printing, Suri 100, Svenman, Svetovid, SwalleyD, Syd1435, Symane, TAS, THEN WHO WAS PHONE?, Tabletop, Tagishsimon, TaintedMustard, Takamaxa, Taksim25, Tangent747, Taoster, Tarquin, Tassedethe, Tatakubuntu, Tcr25, Tedernst, Tellyaddict, Template namespace initialisation script, TerryElliott, Textorus, Tgv8925, The Arbiter, The Wild Falcon, The undertow, TheBourtreehillian, TheEgyptian, TheListUpdater, Theda, Thephotoplayer, ThomasAndrewNimmo, Tide rolls, Tim PF, Timc, Tmorgan147, Toiyabe, Tony1, TonyTheTiger, Train2104, Trekphiler, Ttsalo, Tudorminstrel, Ulric1313, Undead1, Vapmachado, Vcwizard, Vegas949, Villager57, Voyager, Vuo, WOSlinker, Wahiba, Wavelength, Wdfarmer, Webnetprof, Wereon, WiJG?, WikiHead, Wiki Raja, Wikipelli, William Grimes, Wolfkeeper, Wongm, Woodenships, Woohookitty, Wspencer11, Wuhwuzdat, Yelyos, Yossarian, Yvwv, Zhou Yu, Zoney, Zzuuzz, ŠJů, Joвaнв6, 819 anonymous edits

Jesús Carranza, Veracruz *Source*: http://en.wikipedia.org/w/index.php?title=Jes%C3%BAs_Carranza%2C_Veracruz *Contributors*: Dv6057, Ilion2, Kingpin13, Morbidthoughts, NewEnglandYankee, Protonk, Zorydominguez, 8 anonymous edits

Politics of Mexico *Source*: http://en.wikipedia.org/w/index.php?title=Politics_of_Mexico *Contributors*: 0, Abögarp, Acntx, Addshore, Aitias, Alansohn, AmiDaniel, AndreNatas, Angela, Antiedman, Aoa, Asereje, Babbage, Beetstra, Beland, Blueboy96, Bobblehead, Bogey97, Bongwarrior, Branddobbe, Brion VIBBER, Bron27, Bryan Derksen, Bucketsofg, CWii, Calmer Waters, Caltas, Can't sleep, clown will eat me, Canley, Cantus, CarolSpears, Chivista, Ck lostsword, Clandonohuemama, CommonsDelinker, Coolcaesar, Cooliemon, Correogsk, CrazyPhunk, Curufinwe, DO'Neil, DOCTAPUS, Darkstar1st, Davidcannon, Ddye, Delldot, DerHexer, Descendall, Discospinster, Download, Dreadstar, Dsmithsmithy, Dúnadan, EOZyo, Electionworld, EoGuy, Epbr123, Escape Orbit, Fernkes, Fromgermany, GF210, Gaius Cornelius, GheeBern, Giraffedata, Glane23, Globalorg, Goodbone, Goodboner, Goodnightmush, Hajor, HellaNorCal, Herostratus, Hjr, Horstvonludwig, Husond, I dream of horses, Infrogmation, Int21h, Iota, J.delanoy, JHunterJ, JMurphy, Janisterzaj, Jdso4, Jennavecia, John of Reading, Jolomo, JorgeGG, Jose Icaza, Joseph Solis in Australia, Josh0322, Jusdafax, Katalaveno, Katieh5584, Kbdank71, Kiteinthewind, Kman543210, KnowledgeOfSelf, Korossyl, Koyaanis Qatsi, KoyaanisQatsi, Kukini, Kuru, LeaveSleaves, Leszek Jańczuk, LilHelpa, Looxix, Luctec, Lyctc, M P M, Mahanga, Mankara, Maphisto86, Master of Puppets, Mattypattywatty, Maw, Mercy11, Mikeblas, Mummy34, NawlinWiki, Neelix, Nilfanion, Nlu, Oscarthecat, Peterparkersalterego, Phearson, Plm209, PrestonH, R'n'B, RUL3R, Rdkamp, Recognizance, Reconsider the static, RedWolf, Reedy, Res2216firestar, Ric36, Rich Farmbrough, Rjwilmsi, Roadmaster, Ruakh, Ruiz, Sceptre, Scott Ritchie, Seba5618, Sebesta, Semperf, Seth Ilys, Shanes, Sharkface217, ShatterdRose, Siafu, Sietse Snel, Silversink, Skapur, Snigbrook, Snowolf, Spangineer, Stephen G. Brown, Stoweiam, Stunetii, Sun.portal, Tbhotch, Template namespace initialisation script, The Epopt, The Random Editor, The Thing That Should Not Be, Thornecaxa, Tide rolls, Torc2, Trusilver, Vishnava, Vizcarra, Wd9ewk, Wenteng, WhisperToMe, Will Beback, Wimt, Woohookitty, Wrp103, Yakoo, Yakowljew, Youssefsan, Zoicon5, 387 anonymous edits

Ferrocarril Transistmico *Source*: http://en.wikipedia.org/w/index.php?title=Ferrocarril_Transistmico *Contributors*: Aille, Bolivian Unicyclist, Caerwine, Dravecky, Maury Markowitz, NE2, Peter Horn, Tubezone, Wheeltapper, 8 anonymous edits

Ferrocarriles Nacionales de México *Source*: http://en.wikipedia.org/w/index.php?title=Ferrocarriles_Nacionales_de_M%C3%A9xico *Contributors*: Aille, Barticus88, Bolivian Unicyclist, Bruce Hall, Caerwine, Dough4872, Dravecky, Hugo999, Joseph Solis in Australia, Look2See1, Monster4711, NE2, Slambo, Tubezone, 8 anonymous edits

Privatization *Source*: http://en.wikipedia.org/w/index.php?title=Privatization *Contributors*: -- April, -Majestic-, 137.208.3.xxx, 16@r, 172, Active Banana, AdjustShift, Ahkitj, Alex1709, Alsandro, Altenmann, Andrew Levine, Andreworkney, Angela, AnnaP, Arminius, Asws, Audriusa, AvicAWB, BL, Bagatelle, Beagel, Beland, Bensmith8406, Bequw, Beural Inc, Bgoldenberg, Bird, Blackjack3, Bluemoose, Bob A, Bobblewik, Bobrayner, Bogglevit, Bongwarrior, Bovineone, Britannicus, Bugger4stalin, Byelf2007, Bình Giang, CBowers, Cate, Catgut, Childhoodsend, Chrihern, Christopher Kraus, Cimon Avaro, Codwayne, Colchicum, Conversion script, Counterheq, Cretog8, CyberSach, Damnedkingdom, Daniel Quinlan, Ddxc, Defender of torch, Degen Earthfast, DelosHarriman, DerHexer, Dhodges, Dinosaurdarrell, Discospinster, Dodge1884, Dr kapp 89, Dsmccohen, Dvavasour, Ed Poor, Edward, Edward321, El C, Elliotreed, Elnocturno, EqualRights, Ertly, Ethnoid, Evercat, Examtester, Eyedubya, Faizul Latif Chowdhury, Fang 23, Femto, Foodchain, Forp, Fuzheado, G-Man, GeMiJa, Germabel, Glen, Gogo Dodo, Golbez, Goplat, GovernmentAndBusiness, Graham87, Grampion76, Gregalton, Grue, Grz77, Guaka, Gurchzilla, H.ehsaan, H@r@ld, Haham hanuka, Harburg, Hassanerpupone, Hauser, Headbomb, Hede2000, Helix84, Heron, Herschelkrustofsky, Historianscholar, Hongooi, Hroðulf, Hugo999, Hulyad, Humehwy, Ikari, Inwind, Iohannes Animosus, Iplaw, Iridescent, Itpi, JHunterJ, JLMadrigal, Jail stalinazis, Jamyskis, Jandalhandler, Java13690, Jecar, Jeff G., Jenks24, Jerryseinfeld, Jersey Devil, Jhsprague, Jklin, Jnivekk, John Quiggin, JonHarder, Joseph Solis in Australia,

Jossi, Justice for All, Jwolfe7, KDRGibby, Karina.l.k, Keridwen, Khalid hassani, Kieff, Kollision, Larklight, LauraSal, LilHelpa, Lir, Llywrch, Looxix, Lord of the Isles, Lumidek, Lussmu, Mandarax, Marcika, Martha p, MartinHarper, Masao, Materialscientist, MattieTK, Mego'brien, Memo232, Mervyn, Mickyclarke, Midway, Mike Christie, Movementarian, Naddy, Nakon, Nat Krause, Nbarth, Nikodemos, No Guru, Noraft, Northmeister, Nposs, Ohconfucius, Olivier, Oolong, Opihi salad, Orangesodakid, Pasixxxx, Patrick, Pavel Vozenilek, Pcb21, PeterCrispin, PhilLiberty, Phoyer, Piotrus, ProhibitOnions, Ps07swt, Punkche, R Lowry, REDyellowGreenBLUE, RHaden, RM21, Rama, RattleMan, Rd232, Rdkamp, Reedy, Rich Farmbrough, Rich257, Rjecina, Rjwilmsi, Rkitko, Roadrunner, Robert Brook, Rohita, RolandR, Rondack, Rrburke, Rsm99833, Rumping, SDC, Sam Hocevar, Sam Spade, Sconnie, Sekicho, SensiStarToaster, Sergeymk, Sert710, Shiften1981, SimonP, SimonTrew, Simonpage, Singhalawap, SiobhanHansa, Sir Stanley, SlimVirgin, Smack, Smokizzy, Sonderbro, Spangineer, Spikey, Stambolov, Ste1n, Stephenb, Stewacide, Summit84, Summitraj, Sungl, Superborsuk, Susan Mason, Swliv, Syp, TOttenville8, Tagishsimon, TakuyaMurata, Tannin, Tarquin, Taskforce, Tempodivalse, ThaddeusB, The Behnam, Thirdcoastbuck, Tide rolls, Tiles, Tmh, Tomwalden, Treisijs, Trek78, Tri400, Triskaideka, Troopedagain, Tslocum, Ultramarine, Urbanette, Utcursch, Utenzy, Vald, Valois bourbon, Vandalismterminator, Vanished User 8a9b4725f8376, Vaoverland, Versageek, Viajero, Vroman, Vuo, Wadayow, Wavelength, Wdyoung, Welshram89, Wikidea, Will Beback, Wisq, Woohookitty, Wtmitchell, Xerstau, Xxpor, Yaturi3, Yzb, Zarcadia, Zenohockey, Zhou Yu, Zzyzx11, Île flottante, 405 anonymous edits

Ferrosur *Source*: http://en.wikipedia.org/w/index.php?title=Ferrosur *Contributors*: Arturoramos, Bolivian Unicyclist, Caerwine, Cla68, Dough4872, Dravecky, Dvavasour, Hugo999, Jamcib, LGMcM, Liesel, LilHelpa, Lotje, Mackensen, NE2, Oknazevad, Peter Horn, R'son-W, Rjwilmsi, Tabletop, Tubezone, 6 anonymous edits

Image Sources, Licenses and Contributors

File:Décret de nationalisation des locaux de couvent Sainte-Cécile (1793) (Nationalization decree affecting the premises of the convent of Sainte-Cécile), Grenoble, France.jpg *Source*: http://en.wikipedia.org/w/index.php?title=File:Décret_de_nationalisation_des_locaux_de_couvent_Sainte-Cécile_(1793)_(Nationalization_decree_affecting_the_premises_of_the_convent_of_Sainte-Cécile),_Grenoble,_ *License*: unknown *Contributors*: jlreymond

Image:Soo Locks-Sault-Ste Marie.png *Source*: http://en.wikipedia.org/w/index.php?title=File:Soo_Locks-Sault-Ste_Marie.png *License*: unknown *Contributors*: AnRo0002, Appraiser, Feydey, Geofrog, Jkelly, Juiced lemon, Kimdime, Mattes, Mircea, Rmhermen, Skeezix1000, Xnatedawgx, Yassie, 5 anonymous edits

File:Brno, Brno Město, historická koňská tramvaj.jpg *Source*: http://en.wikipedia.org/w/index.php?title=File:Brno,_Brno_Město,_historická_koňská_tramvaj.jpg *License*: unknown *Contributors*: User:Aktron/Nápověda

File:5051 Earl Bathurst Cocklewood Harbour.jpg *Source*: http://en.wikipedia.org/w/index.php?title=File:5051_Earl_Bathurst_Cocklewood_Harbour.jpg *License*: unknown *Contributors*: Alex at kms, Duncharris, Geof Sheppard, Nilfanion, RHaworth, Railwayfan2005, Shortfatlad, 1 anonymous edits

File:Shinkansen-type-0.jpg *Source*: http://en.wikipedia.org/w/index.php?title=File:Shinkansen-type-0.jpg *License*: unknown *Contributors*: Cassiopeia sweet, Hohoho, Morio, , 1 anonymous edits

File:CTA tracks.jpg *Source*: http://en.wikipedia.org/w/index.php?title=File:CTA_tracks.jpg *License*: unknown *Contributors*: User:Dschwen

File:Ireland - Dublin - Tram.jpg *Source*: http://en.wikipedia.org/w/index.php?title=File:Ireland_-_Dublin_-_Tram.jpg *License*: unknown *Contributors*: User:CGPGrey

File:2TE10U Russian Locomotive.jpg *Source*: http://en.wikipedia.org/w/index.php?title=File:2TE10U_Russian_Locomotive.jpg *License*: unknown *Contributors*: User:Anthony Ivanoff

File:HŽ 7123 series DMU (06).JPG *Source*: http://en.wikipedia.org/w/index.php?title=File:HŽ_7123_series_DMU_(06).JPG *License*: unknown *Contributors*: User:Orlovic

File:InterCity2 - passenger car interior.jpg *Source*: http://en.wikipedia.org/w/index.php?title=File:InterCity2_-_passenger_car_interior.jpg *License*: unknown *Contributors*: Jonik at en.wikipedia

File:Wagons 550.jpg *Source*: http://en.wikipedia.org/w/index.php?title=File:Wagons_550.jpg *License*: unknown *Contributors*: Original uploader was G-Man at en.wikipedia

Image:Pöörangud Tartu raudteejaamas.jpg *Source*: http://en.wikipedia.org/w/index.php?title=File:Pöörangud_Tartu_raudteejaamas.jpg *License*: unknown *Contributors*: User:DJ Sturm

Image:CTA loop junction.jpg *Source*: http://en.wikipedia.org/w/index.php?title=File:CTA_loop_junction.jpg *License*: unknown *Contributors*: User:Dschwen

File:European railway map.jpg *Source*: http://en.wikipedia.org/w/index.php?title=File:European_railway_map.jpg *License*: unknown *Contributors*: User:PeterEastern

File:Eastbound over SCB.jpg *Source*: http://en.wikipedia.org/w/index.php?title=File:Eastbound_over_SCB.jpg *License*: unknown *Contributors*: Davepape, Gürbetaler, Iain Bell, Ronaldino, Thryduulf, Voyager, 2 anonymous edits

File:HBD DD1.jpg *Source*: http://en.wikipedia.org/w/index.php?title=File:HBD_DD1.jpg *License*: unknown *Contributors*: User:SwalleyD

File:UP 6670.jpg *Source*: http://en.wikipedia.org/w/index.php?title=File:UP_6670.jpg *License*: unknown *Contributors*: terry cantrell

File:Train wreck at Montparnasse 1895.jpg *Source*: http://en.wikipedia.org/w/index.php?title=File:Train_wreck_at_Montparnasse_1895.jpg *License*: unknown *Contributors*: Studio Lévy and Sons (Studio Lévy & fils)

File:BNSF 5350 20040808 Prairie du Chien WI.jpg *Source*: http://en.wikipedia.org/w/index.php?title=File:BNSF_5350_20040808_Prairie_du_Chien_WI.jpg *License*: unknown *Contributors*: User:Slambo

File:DeutscheBahn gobeirne.jpg *Source*: http://en.wikipedia.org/w/index.php?title=File:DeutscheBahn_gobeirne.jpg *License*: unknown *Contributors*: User:gobeirne

File:Stanhope Station Railway Lines.jpg *Source*: http://en.wikipedia.org/w/index.php?title=File:Stanhope_Station_Railway_Lines.jpg *License*: unknown *Contributors*: Tellyaddict

File:JRE-TEC-E5 omiya.JPG *Source*: http://en.wikipedia.org/w/index.php?title=File:JRE-TEC-E5_omiya.JPG *License*: unknown *Contributors*: User:Toshinori baba

File:Flag of Mexico.svg *Source*: http://en.wikipedia.org/w/index.php?title=File:Flag_of_Mexico.svg *License*: unknown *Contributors*: User:AlexCovarrubias

Image:Archivo General de la Nación.jpg *Source*: http://en.wikipedia.org/w/index.php?title=File:Archivo_General_de_la_Nación.jpg *License*: unknown *Contributors*: Penmexico

File:2006 Mexican election per state.svg *Source*: http://en.wikipedia.org/w/index.php?title=File:2006_Mexican_election_per_state.svg *License*: unknown *Contributors*: Dove, Huhsunqu, Petr Dlouhý, Roberta F., Thelmadatter

File:State governments by party.svg *Source*: http://en.wikipedia.org/w/index.php?title=File:State_governments_by_party.svg *License*: unknown *Contributors*: User:EOZyo

File:Fox133318.jpg *Source*: http://en.wikipedia.org/w/index.php?title=File:Fox133318.jpg *License*: unknown *Contributors*: Bogdan, Carlosar, OsamaK

Image:Ferrosur-map.png *Source*: http://en.wikipedia.org/w/index.php?title=File:Ferrosur-map.png *License*: unknown *Contributors*: User:Liesel

File:FNM 9216 AMALUCAN.jpg *Source*: http://en.wikipedia.org/w/index.php?title=File:FNM_9216_AMALUCAN.jpg *License*: unknown *Contributors*: Iain Bell, Kajuna

File:Nationales de Mexico 1984.jpg *Source*: http://en.wikipedia.org/w/index.php?title=File:Nationales_de_Mexico_1984.jpg *License*: unknown *Contributors*: User:Monster4711

File:NdM Logo Mexico.jpg *Source*: http://en.wikipedia.org/w/index.php?title=File:NdM_Logo_Mexico.jpg *License*: unknown *Contributors*: User:Monster4711

Image:FERROSUR 4400 NORTE.jpg *Source*: http://en.wikipedia.org/w/index.php?title=File:FERROSUR_4400_NORTE.jpg *License*: unknown *Contributors*: Kajuna, Rαge, 3 anonymous edits

File:FSRR FNM 04.jpg *Source*: http://en.wikipedia.org/w/index.php?title=File:FSRR_FNM_04.jpg *License*: unknown *Contributors*: -

GNU Free Documentation License Version 1.2, November 2002 Copyright (C) 2000,2001,2002 Free Software Foundation, Inc. 59 Temple Place, Suite 330, Boston, MA 02111-1307 USA Everyone is permitted to copy and distribute verbatim copies of this license document, but changing it is not allowed.

0. PREAMBLE

The purpose of this License is to make a manual, textbook, or other functional and useful document "free" in the sense of freedom: to assure everyone the effective freedom to copy and redistribute it, with or without modifying it, either commercially or noncommercially. Secondarily, this License preserves for the author and publisher a way to get credit for their work, while not being considered responsible for modifications made by others. This License is a kind of "copyleft", which means that derivative works of the document must themselves be free in the same sense. It complements the GNU General Public License, which is a copyleft license designed for free software. We have designed this License in order to use it for manuals for free software, because free software needs free documentation: a free program should come with manuals providing the same freedoms that the software does. But this License is not limited to software manuals; it can be used for any textual work, regardless of subject matter or whether it is published as a printed book. We recommend this License principally for works whose purpose is instruction or reference.

1. APPLICABILITY AND DEFINITIONS

This License applies to any manual or other work, in any medium, that contains a notice placed by the copyright holder saying it can be distributed under the terms of this License. Such a notice grants a world-wide, royalty-free license, unlimited in duration, to use that work under the conditions stated herein. The "Document", below, refers to any such manual or work. Any member of the public is a licensee, and is addressed as "you". You accept the license if you copy, modify or distribute the work in a way requiring permission under copyright law. A "Modified Version" of the Document means any work containing the Document or a portion of it, either copied verbatim, or with modifications and/or translated into another language. A "Secondary Section" is a named appendix or a front-matter section of the Document that deals exclusively with the relationship of the publishers or authors of the Document to the Document's overall subject (or to related matters) and contains nothing that could fall directly within that overall subject. (Thus, if the Document is in part a textbook of mathematics, a Secondary Section may not explain any mathematics.) The relationship could be a matter of historical connection with the subject or with related matters, or of legal, commercial, philosophical, ethical or political position regarding them. The "Invariant Sections" are certain Secondary Sections whose titles are designated, as being those of Invariant Sections, in the notice that says that the Document is released under this License. If a section does not fit the above definition of Secondary then it is not allowed to be designated as Invariant. The Document may contain zero Invariant Sections. If the Document does not identify any Invariant Sections then there are none. The "Cover Texts" are certain short passages of text that are listed, as Front-Cover Texts or Back-Cover Texts, in the notice that says that the Document is released under this License. A Front-Cover Text may be at most 5 words, and a Back-Cover Text may be at most 25 words. A "Transparent" copy of the Document means a machine-readable copy, represented in a format whose specification is available to the general public, that is suitable for revising the document straightforwardly with generic text editors or (for images composed of pixels) generic paint programs or (for drawings) some widely available drawing editor, and that is suitable for input to text formatters or for automatic translation to a variety of formats suitable for input to text formatters. A copy made in an otherwise Transparent file format whose markup, or absence of markup, has been arranged to thwart or discourage subsequent modification by readers is not Transparent. An image format is not Transparent if used for any substantial amount of text. A copy that is not "Transparent" is called "Opaque". Examples of suitable formats for Transparent copies include plain ASCII without markup, Texinfo input format, LaTeX input format, SGML or XML using a publicly available DTD, and standard-conforming simple HTML, PostScript or PDF designed for human modification. Examples of transparent image formats include PNG, XCF and JPG. Opaque formats include proprietary formats that can be read and edited only by proprietary word processors, SGML or XML for which the DTD and/or processing tools are not generally available, and the machine-generated HTML, PostScript or PDF produced by some word processors for output purposes only. The "Title Page" means, for a printed book, the title page itself, plus such following pages as are needed to hold, legibly, the material this License requires to appear in the title page. For works in formats which do not have any title page as such, "Title Page" means the text near the most prominent appearance of the work's title, preceding the beginning of the body of the text. A section "Entitled XYZ" means a named subunit of the Document whose title either is precisely XYZ or contains XYZ in parentheses following text that translates XYZ in another language. (Here XYZ stands for a specific section name mentioned below, such as "Acknowledgements", "Dedications", "Endorsements", or "History".) To "Preserve the Title" of such a section when you modify the Document means that it remains a section "Entitled XYZ" according to this definition. The Document may include Warranty Disclaimers next to the notice which states that this License applies to the Document. These Warranty Disclaimers are considered to be included by reference in this License, but only as regards disclaiming warranties: any other implication that these Warranty Disclaimers may have is void and has no effect on the meaning of this License.

2. VERBATIM COPYING

You may copy and distribute the Document in any medium, either commercially or noncommercially, provided that this License, the copyright notices, and the license notice saying this License applies to the Document are reproduced in all copies, and that you add no other conditions whatsoever to those of this License. You may not use technical measures to obstruct or control the reading or further copying of the copies you make or distribute. However, you may accept compensation in exchange for copies. If you distribute a large enough number of copies you must also follow the conditions in section 3. You may also lend copies, under the same conditions stated above, and you may publicly display copies.

3. COPYING IN QUANTITY

If you publish printed copies (or copies in media that commonly have printed covers) of the Document, numbering more than 100, and the Document's license notice requires Cover Texts, you must enclose the copies in covers that carry, clearly and legibly, all these Cover Texts: Front-Cover Texts on the front cover, and Back-Cover Texts on the back cover. Both covers must also clearly and legibly identify you as the publisher of these copies. The front cover must present the full title with all words of the title equally prominent and visible. You may add other material on the covers in addition. Copying with changes limited to the covers, as long as they preserve the title of the Document and satisfy these conditions, can be treated as verbatim copying in other respects. If the required texts for either cover are too voluminous to fit legibly, you should put the first ones listed (as many as fit reasonably) on the actual cover, and continue the rest onto adjacent pages. If you publish or distribute Opaque copies of the Document numbering more than 100, you must either include a machine-readable Transparent copy along with each Opaque copy, or state in or with each Opaque copy a computer-network location from which the general network-using public has access to download using public-standard network protocols a complete Transparent copy of the Document, free of added material. If you use the latter option, you must take reasonably prudent steps, when you begin distribution of Opaque copies in quantity, to ensure that this Transparent copy will remain thus accessible at the stated location until at least one year after the last time you distribute an Opaque copy (directly or through your agents or retailers) of that edition to the public. It is requested, but not required, that you contact the authors of the Document well before redistributing any large number of copies, to give them a chance to provide you with an updated version of the Document.

4. MODIFICATIONS

You may copy and distribute a Modified Version of the Document under the conditions of sections 2 and 3 above, provided that you release the Modified Version under precisely this License, with the Modified Version filling the role of the Document, thus licensing distribution and modification of the Modified Version to whoever possesses a copy of it. In addition, you must do these things in the Modified Version: A. Use in the Title Page (and on the covers, if any) a title distinct from that of the Document, and from those of previous versions (which should, if there were any, be listed in the History section of the Document). You may use the same title as a previous version if the original publisher of that version gives permission. B. List on the Title Page, as authors, one or more persons or entities responsible for authorship of the modifications in the Modified Version, together with at least five of the principal authors of the Document (all of its principal authors, if it has fewer than five), unless they release you from this requirement. C. State on the Title page the name of the publisher of the Modified Version, as the publisher. D. Preserve all the copyright notices of the Document. E. Add an appropriate copyright notice for your modifications adjacent to the other copyright notices. F. Include, immediately after the copyright notices, a license notice giving the public permission to use the Modified Version under the terms of this License, in the form shown in the Addendum below. G. Preserve in that license notice the full lists of Invariant Sections and required Cover Texts given in the Document's license notice. H. Include an unaltered copy of this License. I. Preserve the section Entitled "History", Preserve its Title, and add to it an item stating at least the title, year, new authors, and publisher of the Modified Version as given on the Title Page. If there is no section Entitled "History" in the Document, create one stating the title, year, authors, and publisher of the Document as given on its Title Page, then add an item describing the Modified Version as stated in the previous sentence. J. Preserve the network location, if any, given in the Document for public access to a Transparent copy of the Document, and likewise the network locations given in the Document for previous versions it was based on. These may be placed in the "History" section. You may omit a network location for a work that was published at least four years before the Document itself, or if the original publisher of the version it refers to gives permission. K. For any section Entitled "Acknowledgements" or "Dedications", Preserve the Title of the section, and preserve in the section all the substance and tone of each of the contributor acknowledgements and/or dedications given therein. L. Preserve all the Invariant Sections of the Document, unaltered in their text and in their titles. Section numbers or the equivalent are not considered part of the section titles. M. Delete any section Entitled "Endorsements". Such a section may not be included in the Modified Version. N. Do not retitle any existing section to be Entitled "Endorsements" or to conflict in title with any Invariant Section. O. Preserve any Warranty Disclaimers. If the Modified Version includes new front-matter sections or appendices that qualify as Secondary Sections and contain no material copied from the Document, you may at your option designate some or all of these sections as invariant. To do this, add their titles to the list of Invariant Sections in the Modified Version's license notice. These titles must be distinct from any other section titles. You may add a section Entitled "Endorsements", provided it contains nothing but endorsements of your Modified Version by various parties--for example, statements of peer review or that the text has been approved by an organization as the authoritative definition of a standard. You may add a passage of up to five words as a Front-Cover Text, and a passage of up to 25 words as a Back-Cover Text, to the end of the list of Cover Texts in the Modified Version. Only one passage of Front-Cover Text and one of Back-Cover Text may be added by (or through arrangements made by) any one entity. If the Document already includes a cover text for the same cover, previously added by you or by arrangement made by the same entity you are acting on behalf of, you may not add another; but you may replace the old one, on explicit permission from the previous publisher that added the old one. The author(s) and publisher(s) of the Document do not by this License give permission to use their names for publicity for or to assert or imply endorsement of any Modified Version.

5. COMBINING DOCUMENTS

You may combine the Document with other documents released under this License, under the terms defined in section 4 above for modified versions, provided that you include in the combination all of the Invariant Sections of all of the original documents, unmodified, and list them all as Invariant Sections of your combined work in its license notice, and that you preserve all their Warranty Disclaimers. The combined work need only contain one copy of this License, and multiple identical Invariant Sections may be replaced with a single copy. If there are multiple Invariant Sections with the same name but different contents, make the title of each such section unique by adding at the end of it, in parentheses, the name of the original author or publisher of that section if known, or else a unique number. Make the same adjustment to the section titles in the list of Invariant Sections in the license notice of the combined work. In the combination, you must combine any sections Entitled "History" in the various original documents, forming one section Entitled "History"; likewise combine any sections Entitled "Acknowledgements", and any sections Entitled "Dedications". You must delete all sections Entitled "Endorsements".

6. COLLECTIONS OF DOCUMENTS

You may make a collection consisting of the Document and other documents released under this License, and replace the individual copies of this License in the various documents with a single copy that is included in the collection, provided that you follow the rules of this License for verbatim copying of each of the documents in all other respects. You may extract a single document from such a collection, and distribute it individually under this License, provided you insert a copy of this License into the extracted document, and follow this License in all other respects regarding verbatim copying of that document.

7. AGGREGATION WITH INDEPENDENT WORKS

A compilation of the Document or its derivatives with other separate and independent documents or works, in or on a volume of a storage or distribution medium, is called an "aggregate" if the copyright resulting from the compilation is not used to limit the legal rights of the compilation's users beyond what the individual works permit. When the Document is included in an aggregate, this License does not apply to the other works in the aggregate which are not themselves derivative works of the Document. If the Cover Text requirement of section 3 is applicable to these copies of the Document, then if the Document is less than one half of the entire aggregate, the Document's Cover Texts may be placed on covers that bracket the Document within the aggregate, or the electronic equivalent of covers if the Document is in electronic form. Otherwise they must appear on printed covers that bracket the whole aggregate.

8. TRANSLATION

Translation is considered a kind of modification, so you may distribute translations of the Document under the terms of section 4. Replacing Invariant Sections with translations requires special permission from their copyright holders, but you may include translations of some or all Invariant Sections in addition to the original versions of these Invariant Sections. You may include a translation of this License, and all the license notices in the Document, and any Warranty Disclaimers, provided that you also include the original English version of this License and the original versions of those notices and disclaimers. In case of a disagreement between the translation and the original version of this License or a notice or disclaimer, the original version will prevail. If a section in the Document is Entitled "Acknowledgements", "Dedications", or "History", the requirement (section 4) to Preserve its Title (section 1) will typically require changing the actual title.

9. TERMINATION

You may not copy, modify, sublicense, or distribute the Document except as expressly provided for under this License. Any other attempt to copy, modify, sublicense or distribute the Document is void, and will automatically terminate your rights under this License. However, parties who have received copies, or rights, from you under this License will not have their licenses terminated so long as such parties remain in full compliance.

10. FUTURE REVISIONS OF THIS LICENSE

The Free Software Foundation may publish new, revised versions of the GNU Free Documentation License from time to time. Such new versions will be similar in spirit to the present version, but may differ in detail to address new problems or concerns. See http://www.gnu.org/copyleft/. Each version of the License is given a distinguishing version number. If the Document specifies that a particular numbered version of this License "or any later version" applies to it, you have the option of following the terms and conditions either of that specified version or of any later version that has been published (not as a draft) by the Free Software Foundation. If the Document does not specify a version number of this License, you may choose any version ever published (not as a draft) by the Free Software Foundation. ADDENDUM: How to use this License for your documents To use this License in a document you have written, include a copy of the License in the document and put the following copyright and license notices just after the title page: Copyright (c) YEAR YOUR NAME. Permission is granted to copy, distribute and/or modify this document under the terms of the GNU Free Documentation License, Version 1.2 or any later version published by the Free Software Foundation; with no Invariant Sections, no Front-Cover Texts, and no Back-Cover Texts. A copy of the license is included in the section entitled "GNU Free Documentation License". If you have Invariant Sections, Front-Cover Texts and Back-Cover Texts, replace the "with...Texts." line with this: with the Invariant Sections being LIST THEIR TITLES, with the Front-Cover Texts being LIST, and with the Back-Cover Texts being LIST. If you have Invariant Sections without Cover Texts, or some other combination of the three, merge those two alternatives to suit the situation. If your document contains nontrivial examples of program code, we recommend releasing these examples in parallel under your choice of free software license, such as the GNU General Public License, to permit their use in free software.

Printed by Books on Demand GmbH, Norderstedt / Germany